CAMBRIDGE COUNTY GEOGRAPHIES

General Editor: F. H. H. GUILLEMARD, M.A., M.D.

BUCKINGHAMSHIRE

Cambridge County Geographies

BUCKINGHAMSHIRE

by

A. MORLEY DAVIES, D.Sc., F.G.S.

With Maps, Diagrams and Illustrations

Cambridge :
at the University Press
1912

CAMBRIDGE UNIVERSITY PRESS
Cambridge, New York, Melbourne, Madrid, Cape Town,
Singapore, São Paulo, Delhi, Mexico City

Cambridge University Press
The Edinburgh Building, Cambridge CB2 8RU, UK

Published in the United States of America by Cambridge University Press, New York

www.cambridge.org
Information on this title: www.cambridge.org/9781107613584

First published 1912
First paperback edition 2013

A catalogue record for this publication is available from the British Library

ISBN 978-1-107-61358-4 Paperback

PREFACE

THE author is indebted to Messrs Richard Bentley of Slough and E. Mawley of Berkhampstead for the meteorological statistics in Chapter 8, and to Mr Henry J. Turner for the entomological paragraphs in Chapter 7. He must also express his gratitude to Lady Verney for many valuable suggestions and criticisms, which add considerably to whatever value the book may have. To Messrs A. H. Cocks, E. Hollis, W. H. Marsh, W. Niven, C. G. Watkins and others he is also indebted for information on various matters. He must however take upon himself the responsibility for those mistakes which readers with more intimate local knowledge will probably detect, and of which he trusts they will inform him.

A. MORLEY DAVIES.

WINCHMORE HILL,
AMERSHAM.
November, 1911.

CONTENTS

ILLUSTRATIONS

MAPS AND DIAGRAMS

1. County and Shire. The Origin of Buckinghamshire and its Name.

Anyone who thoughtfully considers a map of the counties of central England, of which Buckinghamshire is one, must wonder at the strangeness of their shapes. No king or minister, setting himself to divide up the country into convenient divisions, would produce anything of the sort. The boundaries are neither straight lines like those of so many of the United States or of some of our Australian colonies, nor as a rule do they follow the course of rivers or hill-ranges, except for short distances here and there. Frequently they zigzag across country in the most remarkable fashion.

In the case of a city, the original area of which was once neatly bounded by walls, parts outside the walls may have been added from time to time until a very irregular outline was at last attained ; but there is no reason to suppose that anything similar has happened in the case of counties.

There is plenty of evidence that for centuries our counties remained fixed in shape, and the greatest changes—those of the nineteenth century—have had the effect of simplifying the shape, by abolishing many of the detached parts of one county lying within a neighbouring county, which make the county maps of a hundred years ago still more complicated than those of to-day.

It is evident that these strangely shaped counties were marked out under conditions very unlike any that we meet with nowadays. The map of England shows us that the strangest shapes (and the greatest number of detached parts, if it is an old map) belong to those counties the name of which is made up, as is Bucking-hamshire, by the name of a town followed by the word *shire*. These *shires* were established just a thousand years ago, when Edward the Elder was warring with the Danes. He fortified a number of places against their attacks, and laid an obligation on the neighbouring country to supply men to garrison these forts. The area from which men were drawn for the fort of Buckingham became Buckingham*shire*—the word *shire* meaning a *share*, a division, from an Anglo-Saxon root meaning to cut. What probably really happened is this. A thousand years ago, the freemen from a number of agricultural villages used to meet together once or twice a month at some great tree or other well-known point to discuss public business. This gathering was called the *hundred moot* or *court*, and the group of villages from which the men came to it was called a *hundred*, because it was supposed that there were always a hundred families represented. Through the hundred courts the king levied his taxes, and he treated the hundred as the unit for purposes of taxation, military service, and the like.

When fortresses like Buckingham were to be manned, it was done by allotting to each fortress a certain number of hundreds—in the case of Buckingham, eighteen. The area of these eighteen hundreds became Buckinghamshire. Each hundred consisted of some ten to twenty villages of average size, and the area of land belonging to each village was just what suited it for ploughing, pasture, meadow, and woodland. All the villages in a hundred were not always contiguous; so the shape of the hundred was determined quite accidentally, except where a wide river or high hill-range naturally drove people on opposite sides to attend different courts. Then, when the hundreds were grouped into shires, the boundaries were for the most part equally accidental, and contrasted strongly with those of counties like Kent or Essex, the boundaries of which were boundaries between kingdoms, settled in warfare or by treaty, and therefore comparatively simple lines.

The word *county* is the Norman word (*comté*) corresponding to the English word *shire*, but it is not a translation of it. The corresponding English word would be *earldom*. Before the Norman Conquest it was a general rule that every county had its ruler, who was an *eorl* or earl, and under him a manager or *reeve*, the shire-reeve, or as we now call him the *sheriff*. The Normans replaced the word *eorl* by the Latin word *comes* (meaning a companion) and the shire or earldom became the *comitatus*, which in time became shortened to *county*.

Although Buckinghamshire takes its name from the

town which was originally the county-town, that town is far in the north of the county and difficult of access from the south. Consequently, in course of time, it came about that Aylesbury, which was situated much more centrally and was accessible by main roads from all parts, took its place as the county-town. Nevertheless the county keeps the name of its original county-town, and that in turn, according to some authorities, may have taken its name from the family that first settled there. *Boch* may have been the name of its founder, his sons were called Bochingas, and their home or *ham* was Boch-inga-ham. Thus Buckinghamshire perhaps means "the division belonging to the home of the sons of Boch." Another theory would associate the name with the Anglo-Saxon *boc*, a beech-tree ; but beech-trees characterise South Bucks rather than North, and are rare in that part of the county in which lies the town which gave its name to the shire.

The name "Buckinghamshire" being very long to write, the clerks of the king's exchequer in the middle ages used to abbreviate it by writing only the first four letters and the letter *s* (for shire). From being written thus short, it came to be called "Bucks," and now it is as often spoken of by that name as by its proper title.

2. General Characteristics. Position and Natural Conditions.

Buckinghamshire is an inland county, and only touches its two main rivers, the Thames and the Ouse, far above the lowest points where they are bridged, so that it is cut off from all trade by sea. It is a long county and lies athwart the natural grain of the land, so that it includes a variety of different kinds of land; but they all have this in common, that no coal, or iron ore, or other important minerals are found in any of them (though we cannot yet say what may not be hidden at depths not yet explored). Consequently Buckinghamshire is neither a commercial nor a mining county. Having neither coal nor any important amount of water-power, it cannot be to any great extent a manufacturing county. Mainly therefore it remains, what all counties were in early times, an agricultural county. The raising of crops and cattle, fruit and timber, occupies the greater number of its inhabitants.

In all agricultural districts there are many industries connected with agriculture that are carried on on a small scale, and these are found here as elsewhere. The development of industry on a large scale, to supply the wants of distant places as well as those close at hand, can only take place if there are either (*a*) important local sources of power, such as coal or falling water, or (*b*) a large supply of the raw material necessary for some manufacture, or (*c*) a very large population requiring

certain things to be manufactured. Of these Buckinghamshire has of the first only a little water-power, and
as it has no sea-port it has of the second only what its
own soil produces. It has, however, the huge population
of London near at hand. Consequently some industries
such as the chair-making of Wycombe, dependent
originally on the abundant local timber, have developed
on a fairly large scale.

The nearness of London has other effects on the
county. Several of the main roads and railways from
London to the most thickly populated parts of England
pass through it. Consequently some part of the population makes its living by attending to the wants of
travellers, whether by building railway-coaches at
Wolverton, or boats along the Thames, or by keeping
inns or lodging-houses. Of recent years, too, the corner
of the county nearest to London has become sprinkled
with the houses of persons who have business in London
and travel to and fro every morning and evening.

Farther away, however, the county is as quiet as if
it were a hundred miles from London. Only along the
main roads are motor-cars and bicycles more frequent.
But the foot-passenger can wander along field-paths and
through woods without meeting anyone for hours; for
though the scenery of the county attracts many to do
this, there is ample room for them to scatter themselves
over its area. As Buckinghamshire has no hills reaching
a height much over 800 feet, it has no mountain scenery,
but its great attractiveness lies in its wooded character.
From the Chiltern Hills south-eastwards to the Thames,

Village Green, Quainton

(A typical agricultural village in the Vale of Aylesbury. The steps and broken shaft of the ancient village cross are seen in the centre)

beech-woods abound. On parts of the great plain of
central and north Bucks there are oak-woods. These are
remnants in part of the two great forests of Chiltern
and of Bernwood, but added to by extensive modern
plantations.

3. Size. Shape. Boundaries.

Of the 52 counties of England and Wales, no fewer
than 33 are larger than Buckinghamshire, so that it must
be regarded as a small county. Its greatest length, from
the Thames near Staines to Three Shire Wood, where it
meets Northamptonshire and Bedfordshire, on the watershed between the Ouse and the Nene, is about 52 miles.
Its breadth varies in general between 10 and 20 miles.

The area of the county, as it was until early in the
nineteenth century, included one parish, Caversfield,
quite detached from the rest of the county, and did not
include the area of several parishes or parts of parishes
lying within the county but belonging to Oxfordshire or
Hertfordshire. During the reign of William IV, these
"detached parts" were transferred to the county in which
they actually lay, and the compact counties so produced
are now called (rather inaccurately) the "ancient counties."
The ancient county of Buckingham has an area of
475,682 acres, or about 743 square miles. Of late years
various other exchanges have been made along the
boundaries between county and county. Consequently
the present "administrative county," that is to say the

area under the authority of the Buckinghamshire County
Council, contains 479,358 acres, or nearly 749 square
miles.

The boundaries of the county for reasons explained
in Chapter 1, are very complicated, though here and
there, for shorter and longer distances, they follow natural
features such as rivers, streams, or (approximately) watersheds, or artificial roads. If we start at Three Shire
Wood in the north, and follow the boundary westward,
we find that it keeps fairly well along the watershed
between the Ouse and the Nene, through the wooded
region of Yardley Chase and Salcey Forest, until it runs
down to the river Tove at Bozenham Mill. It then
follows this river southwards until it joins the Ouse near
Wolverton, and then continues up the Ouse to Thornton.
Here it turns north, and follows an altogether irregular
line through Whittlewood Forest to reach the Ouse again
at Biddlesden (eight miles in a straight line from where
it left it). It now follows *down* the Ouse, in a semi-
circular course, to Water Stratford near the town of
Buckingham. During this part of its course, Northamptonshire is left and Oxfordshire becomes the
neighbouring county.

From this point the boundary follows a perfectly
straight course for two miles in a south-westerly direction.
The last part of this is along a Roman road, still in use,
and there can be no doubt that the remainder also originally
followed the road, though that has now disappeared. The
name of Water Stratford (the street-ford) shows that there
the road crossed the river.

Leaving the Roman road near Finmere railway-station, it takes again a very irregular course through what was once the Forest of Bernwood, sometimes following small stream-courses, lanes, and field-boundaries, cutting Akeman Street just north of Piddington, and at one point crossing the prominent Muswell Hill, where an ancient camp looks out across the great plain of Oxfordshire—until it reaches the river Thame at Worminghall, 17 miles nearly due south of the point where it left the Ouse.

For six miles it follows the Thame to the town of that name, and then its tributary the Ford Brook, eastwards to Aston Sandford Manor. Then it takes again a very irregular course southwards to the Thames at Fawley Court, near Henley, on its way crossing the Chiltern Hills at 800 feet above sea-level. For nearly 30 miles of its sinuous course the Thames is the southern boundary of the county, separating it from Berkshire ; though the actual boundary is sometimes in mid-stream, sometimes on the left (or northern) bank. For the last two or three miles, Surrey is the adjoining county.

Near Staines, the boundary leaves the Thames, the Colne here separating Buckinghamshire from Middlesex. The Colne is a river which has changed its course from time to time, and some of these changes must have taken place since the county boundary was settled, as it follows the old course instead of the new. At a point 11 or 12 miles north of the Thames, a bit of Hertfordshire wedges itself in between Buckinghamshire and Middlesex, and pushes the boundary away from the Colne, to pursue

Cliveden Reach

(The steep and wooded Buckinghamshire bank contrasts with the low meadowed Berkshire bank)

once more an extraordinarily irregular course, following ridges, streams, and roads for short distances and leaving them again, until, after crossing the Chilterns three times, backwards and forwards, it reaches the river Ouzel, which it follows, with Bedfordshire on the opposite bank, to Leighton Buzzard. Another long irregular portion follows, ending on the Ouse at Newton Blossomville. Two miles down the Ouse and a straight cut across country bring us to the starting-point at Three Shire Wood.

One peculiar and rather regrettable feature will strike anyone who thus follows the margin of the county—the number of market-towns that lie just outside the boundary, and the few that lie just inside. Brackley in Northamptonshire, Thame and Henley in Oxfordshire, Maidenhead and Windsor in Berkshire, Uxbridge in Middlesex, Rickmansworth and Tring in Hertfordshire, Leighton Buzzard in Bedfordshire, all attract to themselves much business from the adjacent Buckinghamshire villages and so tend to make the population and wealth of our county less than it might be. There is no town within the county so near its margin as to attract the inhabitants of other counties to the same extent, though Olney, Fenny and Stony Stratford, Buckingham, Aylesbury, and Marlow do so in a small degree.

The tracing of the boundaries, whether of a county or of a parish, on the actual ground, is an interesting and instructive occupation. Formerly it was a duty undertaken by the inhabitants of every parish either once a year or every three years, on Ascension Day to walk all round the parish, visiting every stone, tree, or other object that served as a boundary-mark. This was known as "beating the bounds," and though this custom has been rendered unnecessary by the making of Ordnance maps, and has generally fallen into disuse, it would be well if our local schools made some attempt to revive it as part of the teaching of geography and history. At least we may hope that every one who reads this book will try to learn what are the boundaries of his own parish, and of his county if he lives near its margin.

4. Surface and General Features.

Buckinghamshire is divided into two unlike parts by the crest of the Chiltern Hills, which run across the county from south-west to north-east. Standing at some high point on this ridge, such as the monument on Coombe Hill, Wendover, and looking to the north-west, we see the ground at our feet falling steeply down for about 500 feet, and then gradually sloping off to a great plain which extends as far as the eye can reach. This steep slope, from its resemblance to the artificial outer slope of a fortification, is called the Chiltern escarpment (see pp. 14 and 36). The scattered hills on the plain nowhere exceed 700 feet above sea-level, and even the watersheds between the Thames and the Ouse basins, and between those of the Ouse and Nene, rarely rise above 500 feet in height.

The soil of this great plain is for the most part

clayey. In wet weather it soon gets muddy, while in very dry summers it shrinks, and great cracks an inch or more broad and several feet in depth may be found in it. Ditches usually contain water, small streams and marshes are everywhere common, and the larger streams, such as the Thame and the Ouse, are in most places

Pulpit Hill, Kimble
(*Showing the profile of the Chiltern escarpment*)

bordered by flat, low meadows which in rainy seasons are often flooded. The fields are much more often pastures for cattle or meadows for hay than ploughed fields with crops.

At one time, however, this was not the case. We may notice that the pastures and meadows nearly always show a remarkable appearance, being crossed by long, parallel, slightly curved ridges, alternating with hollows. These ridges, from the middle of one hollow to the middle of the next, are about 30 feet wide ; and they rise about 18 inches above the hollows, so that this sort of ground is very tiring to walk across.   That these ridges

The Plain of North Bucks, seen from the Chiltern Hills

and furrows are ancient is shown by the fact that they are sometimes crossed by the present field-boundaries, and may even be seen continuing on the opposite side of a road.

These ridges or "lands" were produced by an ancient method of ploughing, not now practised in Buckinghamshire, though still, or within recent years, in use in Hampshire; the object being to secure drainage of the clay soil, in times when drain-pipes were unknown. They show that great areas of the North Buckinghamshire plain that are now grass-land were once plough-land. Now, instead of providing corn for the inhabitants, it largely provides milk for Londoners.

The houses in this great plain are generally grouped

Grass-land near Aylesbury, showing ancient plough-ridges

in large villages, each with an ancient parish church, showing the village to be of very old date; while here and there we find a church without the village which once accompanied it, as at Fleet Marston, Hillesden, and Thornton; or the church may be in ruins, as at Quarrendon; or have disappeared altogether, as at Hogshaw—the parish having been depopulated in the fifteenth or sixteenth century, when so much land was given up to sheep-runs. All the old cottages in this region have thatched roofs.

If we now turn the other way from the Chiltern crest, we find a very different country. There is no

Characteristic North Bucks Cottages at Botolph Claydon
(Brick and timber; thatched roof)

steep slope to a plain in the south-eastern direction. The ground remains high in general and fairly flat; it is, in fact, a plateau, but it is so much cut up by steep-sided valleys that it is not often recognised as such. There are roads, however, at a height of 500 feet or

more above the sea, along which the cyclist can ride for
miles without dismounting ; and when he comes to a
steep descent, if his road keeps straight, he will soon have
to climb again as high as he went down.   The soil on
this plateau is often clayey or gravelly, but sometimes on

Characteristic South Bucks Cottage, near Amersham

(*Eighteenth century ; brick and flint ; tiled roof.   On the left of the
tree-trunk is seen a little fireside window*)

the high level, and always on the slope of the valley-sides,
it is chalky, and everywhere full of large flints.

Another striking feature of the plateau is its dryness.
There are ponds, fed by rain, dew, and mist, but few
springs and hardly any streams (Map, p. 20).   Even the
valleys are in most cases dry, the chalk which underlies

them readily absorbing all the rain. It is only in the deepest valleys that streams are found, and these, while small in proportion to the size of their valleys, are usually fair-sized from their first appearance, a fact which shows that several tributary streams must have united underground to produce them (see p. 28). The dry valleys are called "bottoms," e.g. Hampden Bottom, Pednor Bottom, Marrod's Bottom; while between them are "ridges" such as Hawridge, Ashridge, Charteridge, etc.

In spite of its dry and stony soil, this Chiltern region is well cultivated. Much corn is grown, and nowhere does the grass-land show signs of having once been ploughed. Instead of the few oak-woods of the plain, there are many beech-woods, and occasionally, where the soil is sandy or gravelly, fir-plantations. A thatched cottage is very difficult to find, all the oldest being tiled, probably because the clays that cover the chalk are very suitable for tile-making, and have long been used for that purpose.

Taking the county as a whole, the feature which marks it off from its neighbours is its abundant woods. In proportion to its area, it has half as much woodland again as any of its neighbours north of the Thames. To find its equal we must go as far west as Gloucestershire. It has also more bare down and heath than its neighbours. But in neither of these respects does it approach the counties south of the Thames.

A few of the best view-points for a survey of the county may be mentioned. The contrast between the Vale and the Chiltern plateau may best be realised from

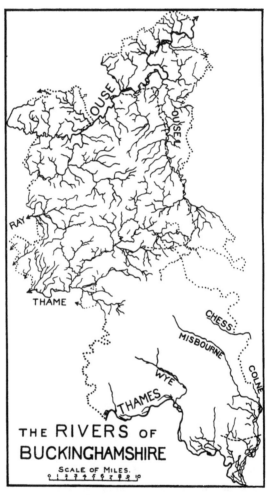

The Rivers of Buckinghamshire

(The contrast between the northern clay plain and southern chalk plateau is well shown. The lines of dashes represent canals)

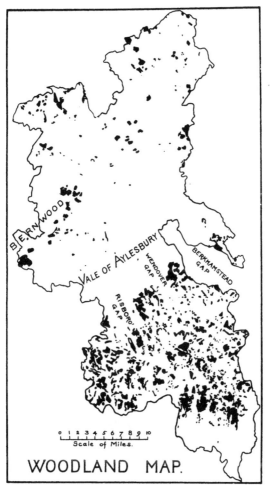

The Woodlands of Buckinghamshire

*(Comparing this with the River Map it will be seen that
where there is most wood there is least water)*

any high point on the Chiltern crest, and of such points there are many, the monument on Coombe Hill near Wendover being perhaps the best. For a good view of the Chiltern escarpment and of the Vale, there are several high points on the hills and ridges in the latter—the old windmill at Stone (near the County Asylum), the summit of Quainton Hill, the high ground at Oving. A walk round the village of Brill, keeping as high up as possible, gives a succession of distant views in all directions, which enable a large area of Buckinghamshire and Oxfordshire to be surveyed. From Bow Brickhill the Ouse valley of North Bucks is well seen. The views in the Chiltern plateau are more restricted as a rule, showing only parts of one valley and the neighbouring high ground. Perhaps the finest is from Beacon Hill, Penn, from which Marlow and the Thames valley with part of the tributary Wye valley may be traced, while the sight ranges far beyond the limits of our county to the South London heights (marked by the Crystal Palace), the Surrey Downs, Windsor Castle, Hindhead, and the Berkshire Downs.

5. Rivers and Watersheds.

The greater part of the water that flows over the surface of Buckinghamshire finds its way finally to the Thames, though for the most part it does not reach that river within the county. The rest makes its way to the Ouse. It might perhaps be expected that the highest hills in the county would separate the waters flowing to such different ends, but this is not the case. From both slopes of the Chilterns the streams are gathered into the Thames, and this is also the case with the highest hills within the northern plain, those around Brill.

On the other hand the watershed between the Thames and the Ouse only attains a greater height than 600 feet above sea-level at one point—Quainton Hill; and for much of its course is below or just above 300 feet. Similarly, the part of the watershed between the Ouse and the Nene that forms part of the county-boundary only exceeds 500 feet above the sea at two or three points, near Luffield Abbey.

The river Ouse is known as the Great Ouse or Bedford Ouse, to distinguish it from other rivers that bear the same name—derived from an ancient Celtic word meaning *water*, a word that appears in other modifications in the river-names *Usk* and *Esk* (not less, it may be added, in the apparently very different word *whiskey*). This river has its source in Northamptonshire, but after flowing a few miles and increasing in volume by the addition of several small tributaries, it forms the county-boundary of Buckinghamshire during a semi-circular course of some seven or eight miles. During this distance it serves as the means of driving several mills. Now entering right into the county, after a few miles' meandering it reaches the town of Buckingham, which was for centuries the county-town, until the inconvenience of its situation, so far away from the southern part of the county, led to its supersession by Aylesbury.

Beyond Buckingham, the Ouse valley becomes much wider. In about four miles it again becomes the county-boundary, though now Buckinghamshire is on its right bank: in another four it is crossed by the Roman road, still in use as a great highway, known as Watling Street. This now crosses the river and its water-meadows by a long bridge of many arches ; but originally travellers had to ford the river. The place where the *street* crossed by a *ford* came to be called Stratford, and it was specially called Stony Stratford, to distinguish it from Water Stratford and Fenny Stratford, at both of which also a Roman road crossed a river.

A mile or two below Stony Stratford the Ouse receives its only important northern or left-bank tributary within the county—the Tove, up which the county-boundary here runs, so that the Ouse for some distance has again Buckinghamshire banks on each side. It now makes a series of long meandering loops within a wide valley, now nearing the higher ground on one side, and now that on the other, and so to Newport Pagnell, a small market-town at the confluence with the Ouzel, a stream draining down from the Chilterns about Ivinghoe. The name of Newport has no reference to shipping, the word *port* in old times meaning a market. From Newport Pagnell, the river continues its winding course northwards to Olney, and then after again serving as county-boundary for a short distance, but this time with Buckinghamshire on its *left* bank, it leaves it altogether.

The river Thame is one of the left-bank or northern tributaries of the Thames. Its source and confluence are both outside the county. It rises in the Chilterns, in that part of Hertfordshire which runs like a blunt wedge into the eastern portion of Buckinghamshire. A large number of tributaries soon join to make it a fair-sized stream as it flows through the Vale of Aylesbury. It is bordered on each side by an area of water-meadows,

Bridge over the Ouse at Olney

(Its "wearisome but necessary length" indicates the width of the river in flood)

liable to flood; and therefore has no towns or villages on its banks until it enters a narrow part of its valley, where close to its right bank stand the little villages of Lower Winchendon and Chearsley, and a little farther on the ruins of Notley Abbey. As it approaches

The Thame near Lower Winchendon

(Here the ground is seen to rise quickly away from the river, out of the reach of floods)

the town which bears its name, the Thame becomes the boundary between Buckinghamshire and Oxfordshire, and continues as such for some miles, until, near Waterperry, it leaves the former county altogether.

The streams that run in a south-easterly direction, from the Chiltern Hills towards the Thames, occupy valleys cut deep below the general surface of the Chiltern plateau. On ascending to the source of one of these streams—the Wye, the Misbourne, or the Chess—it is found that the valley does not end there, but is continued far higher as a dry valley with dry tributary-valleys. In the case of the first two of these rivers, the dry valley extends right up to the Chiltern escarpment, and ends in a deep notch in the line of hills. These notches or gaps are valuable as routes for roads and railways across the Chilterns. They lead geologists to the inference that the rivers were once bigger and rose higher up in their valleys, even beyond what is now the Chiltern watershed. Moreover, as the streams are fair-sized from their very source, it is reasonable to suppose that several small streams have already united into one in their underground course. These streams are large enough to work a number of mills, and the busy little towns of Wycombe and Chesham owe their existence in the first place to their water-power.

The Wye flows direct to the Thames; the Misbourne and Chess enter the Colne, a tributary of the Thames which forms for some distance the eastern boundary of the county. The Chess—which itself forms the boundary for a little distance before passing outside our county to its

Waterside, Chesham

(*Shows the river Chess only a mile below its source. The water, however, is held up by a weir*)

junction at Rickmansworth—has a wide valley, the river
being bordered by low ground scarcely above its level. As
is usual with a slow-flowing river bordered by such alluvial
flats, it several times divides into branches which re-unite
again lower down, while artificial waterways have been
easily made on the low, level ground.   Hence in crossing

Bell Weir, on the Thames, near Wraysbury

the Colne valley by road or railway, three or four
bridges may have to be passed.

Glancing over the streams of the whole county we
may notice some negative characters.   There are no
swift-flowing mountain-brooks, no waterfalls except the
artificial weirs and mill-dams : the streams, large and
small, are all placid, except after very heavy rainfall ; that
is to say the slope or gradient of their bed is very gentle.

In the hilly districts of the Chilterns, where the valley-slopes are steep, no water flows above ground. Steep-sided gorges, too, are almost unknown, but there is one, the Lyde at Bledlow, where a little tributary of the Thame issues from the chalk.

The Thames itself, during the 30 miles in which it forms the county-boundary, is a fine broad river, a

Bell Weir Lock, on the Thames

highway used more for pleasure than for commerce. In the summer, it is full of rowing-boats and house-boats, and many holiday-residences are built along its banks. The reach between Hedsor and Taplow, in particular, is a favourite resort for boating, on account of the beautiful mass of woods on the steep bank (almost a cliff) on the Buckinghamshire side (see p. 11).

For about a century the navigability of the Thames has been increased by "canalisation," that is, the construction of weirs and locks at intervals. A weir is an artificial waterfall, and the effect of making weirs is to change a river from an inclined plane to a sort of liquid staircase, level reaches alternating with sudden falls. Alongside each weir, a lock is constructed to allow boats to pass both upstream and down. In the Buckinghamshire part of the Thames there are nine weirs and locks. In the same distance there are also eight bridges, four for roads and four for railways. The most famous of these probably is Brunel's fine brick bridge at Maidenhead, which carries the Great Western Railway on two high elliptical arches. It is the bridge represented in Turner's famous picture in the National Gallery, "*Rain, Steam, and Speed.*"

In the map on p. 20, an attempt has been made to show every course of running water in the county.

6. Geology and Soil.

By the *Geology* of a district we mean the characters and arrangement of the rocks that underlie its soil—the word *rock* being taken to include the softer masses of clay, sand, and gravel, as well as the harder masses to which it is specially applied in ordinary language. Everybody knows that the rocks immediately below the soil differ from place to place even within a small area, for in one place he may see a chalk-pit, in another a gravel-pit, and not far off the soil may be so heavy as to indicate that there is clay beneath. Quarrymen, too, are familiar with the arrangement of rocks in layers, beds, or courses, and know that in some places these lie horizontally while elsewhere they slope or *dip* so that one passes below another down into the earth. It is by putting together all this local knowledge and adding to it what has been found from deep well-sinkings and mines, that geologists are able to form a pretty accurate idea of the arrangement of the rocks throughout the country down to a depth of thousands of feet.

Besides the rocks already mentioned as being arranged in layers, or *strata*, and accordingly known as stratified rocks, there are others which show no such arrangement. These are known as igneous rocks, and though we have none such anywhere near the surface in Buckinghamshire, examples of them may be seen in some of the "granites" that are brought into the county from Leicestershire or elsewhere for use in repairing the roads.

By tracing the different layers of stratified rock all over the country, and observing how one dips or passes under another, it has been possible to divide these rocks into groups and systems. From their arrangement in layers, from some of them being made up of fragments broken off other rocks, and from most of them containing the fossil remains of dead animals and plants, it is inferred that these rocks were formed by the deposit of layers of sand, mud, and so forth under water, at the bottom of seas or lakes or rivers. From the fact that different groups contain different kinds of fossils, and that those

	NAMES OF SYSTEMS	SUBDIVISIONS	CHARACTERS OF ROCKS
TERTIARY	**Recent** **Pleistocene***	Metal Age Deposits Neolithic ,, Palaeolithic ,, Glacial ,,	Superficial Deposits
	Pliocene	Cromer Series Weybourne Crag Chillesford and Norwich Crags Red and Walton Crags Coralline Crag	Sands chiefly
	Miocene	Absent from Britain	
	Oligocene	Fluviomarine Beds of Hampshire	
	Eocene*	Bagshot Beds London Clay Oldhaven Beds, Woolwich and Reading Thanet Sands [Groups	Clays and Sands chiefly
SECONDARY	**Cretaceous***	Chalk Upper Greensand and Gault Lower Greensand Weald Clay Hastings Sands	Chalk at top Sandstones, Mud and Clays below
	Jurassic*	Purbeck Beds Portland Beds Kimmeridge Clay Corallian Beds Oxford Clay and Kellaways Rock Cornbrash Forest Marble Great Oolite with Stonesfield Slate Inferior Oolite Lias—Upper, Middle, and Lower	Shales, Sandstones and Oolitic Limestones
	Triassic	Rhaetic Keuper Marls Keuper Sandstone Upper Bunter Sandstone Bunter Pebble Beds Lower Bunter Sandstone	Red Sandstones and Marls, Gypsum and Salt
PRIMARY	**Permian**	Magnesian Limestone and Sandstone Marl Slate Lower Permian Sandstone	Red Sandstones and Magnesian Limestone
	Carboniferous	Coal Measures Millstone Grit Mountain Limestone Basal Carboniferous Rocks	Sandstones, Shales and Coals at top Sandstones in middle Limestone and Shales below
	Devonian	Upper Mid Devonian and Old Red Sand- Lower stone	Red Sandstones, Shales, Slates and Lime- stones
	Silurian	Ludlow Beds Wenlock Beds Llandovery Beds	Sandstones, Shales and Thin Limestones
	Ordovician	Caradoc Beds Llandeilo Beds Arenig Beds	Shales, Slates, Sandstones and Thin Limestones
	Cambrian	Tremadoc Slates Lingula Flags Menevian Beds Harlech Grits and Llanberis Slates	Slates and Sandstones
	Pre-Cambrian	No definite classification yet made	Sandstones, Slates and Volcanic Rocks

* These systems outcrop in Buckinghamshire.

which dip down below all the others contain the fossils
least like the animals and plants of the present day, while
those which lie above all others have fossils resembling
animals and plants now living, it is concluded that the
different rock-groups were formed at far-distant periods
of time. The table on p. 33 shows the chief divisions
of the stratified rocks, those placed at the bottom being
the oldest—that is to say that they were formed first,
and may be found underneath any of the others.
The rest follow in order of age, the youngest being
at the top. If at any place in the country the rocks
of any particular system come to the surface (just below
the soil), we know that by sinking a well through
them we *may* come to any of those whose names are
below it on the list, but we shall not come to any whose
names are above it.

The surface rocks of Buckinghamshire belong to the
Jurassic, Cretaceous, Eocene, and Pleistocene systems only.
There probably are rocks of older systems deep beneath
these underground, but the search for these within the
county is only beginning.

The two thickest and most important rock-layers are
(1) clay of Jurassic age, 600 or 700 feet thick, and
(2) chalk of Cretaceous age, about 800 or more feet
thick. Below the first of these, between the two, and
above the second, are three series of rocks, much thinner
and much more variable in character. All these beds
lie with a gentle inclination or *dip* towards the south-
east, so that if we cross the county from north to south
we meet with newer and newer beds in regular order,

the older one by one disappearing under the next newer. Thus the Chalk, which rises up to 800 feet above the sea in the Chilterns, is down at river-level in the Colne valley.

This regular succession of beds is the case only with the Jurassic, Cretaceous, and Eocene strata. The Pleistocene beds are spread in patches all over the county and may rest upon any of the older rocks.

The two thick masses of clay and chalk give rise, by their differences of character, to the differences of soil and scenery that distinguish the two main parts of the county ; while the three variable series of beds that alternate with them, and the equally variable Pleistocene deposits that are scattered over all alike, give rise to the minor diversities of soil and scenery that break the monotony there would otherwise be.

Beginning at the north of the county with the oldest rocks, the following is the succession of beds below the great clay-series (the lowest or oldest being placed at the bottom of the list) :—

	NAME	CHARACTER	THICKNESS IN FEET
LOWER OOLITES	Cornbrash	Rubbly earthy limestone	10 or less
	Great Oolite Clay (with Forest Marble)	Marly clay with thin bands of limestone	15 or less
	Great Oolite Limestone	White and grey limestones	20 or less
	Upper Estuarine Series	Clays and sands	20 or less
	Lower Estuarine Series	Clays and sands	10 to 0
	Northampton Sands	Brown sand and sandstone	5 to 0
	Upper Lias Clay	Grey clay	only the top portion seen

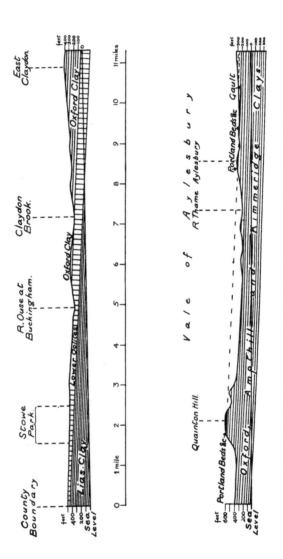

Geological Section across Buckinghamshire from N.N.W. to S.S.E.

(*The scale of height is six times the horizontal scale*)

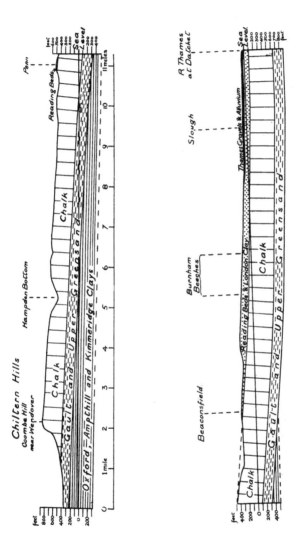

Chiltern Hills

Coombe Hill near Wendover · Hampden Bottom · Reading Beds · Penn

Chalk · Gault and Upper Greensand · Oxford=Ampthill and Kimmeridge Clays

Geological Section across Buckinghamshire—*continued*

Beaconsfield · Burnham Beeches · Slough · R Thames at Datchet

Chalk · Reading Beds & London Clay · Thames Gravels & Alluvium · Gault and Upper Greensand

(*The right-hand end of each of the three upper sections adjoins the left-hand end of the section next below it*)

In the section on pp. 36—37 these strata are collectively named "Lower Oolites." Most of them contain fossils. These, when carefully studied, can teach us much about the geographical conditions which existed in the place where they are found at the time when they were alive. In this case they teach us that North Bucks was, at the beginning and end of the time when these rocks were formed, the bottom of a sea of no great depth, with land not far away ; and in the middle of the time, part of the estuary of some great river. The sand, mud, and other deposits on the sea-bottom or in the estuary are now hardened into these rocks, and the dead bodies of some of the water-animals that lived at that time are preserved as fossils.

These rocks are not of much economic value : the Great Oolite limestone is quarried to some extent for building-stone and lime, and some of the sands and clays are dug, the latter for brick-making, but only to a small extent. The soil which forms at the surface from the breaking up of the rocks and their mixture with the decayed remains of plants, is of a mixed and variable character, and therefore fertile.

The clays that form the sub-soil of the great plain of North Bucks have in them occasional thin beds of limestone, or sometimes other kinds of rock, but these have little influence on the surface-features of the ground. Occasionally large rounded masses of stone are found, which on being broken open show a set of radiating fissures which have been partly or completely filled with crystals of beautifully translucent carbonate of calcium :

Quarry in Great Oolite, Stoke Goldington

such stones are called septaria. Fossils are found in these clays, and show not only that they were laid down at the bottom of the sea, but that the sea had by this time become much deeper than previously. But they teach us more than this, for the fossils found in different layers of the clay are of different kinds, and show that the time during which these clays were accumulating was so long that the animals living in the sea—the marine fauna in scientific language—changed over and over again. It is possible to divide this clay-period into ages by means of the difference in the fossils, and the ages recognised here can be recognised again in other countries. Without going into too much detail, it is sufficient to say that there are three main divisions in the clay—

Kimmeridge Clay,
Ampthill Clay,
Oxford Clay.

The Oxford and Kimmeridge clays have been dug at various places in the county for the making of bricks, but until recently no cutting into the Ampthill clay was known, but it has now been well exposed in deep railway-cuttings. The brick-making industry has become of greater importance in recent years, bricks being now made for house-building in the London district.

The soil above the clay is everywhere wet and tenacious, suitable for pasture and oak-woods rather than for ploughing or for houses to be built upon. Most of the villages in the great clay-plain are situated at spots where newer deposits rest upon the clay, especially where these are gravelly or sandy and so yield a supply of water from springs and wells. Some of these newer deposits, especially in the northern part of the plain, are Pleistocene gravels (see below) ; but in the southern part there are a number of hills formed of the next succeeding series of beds which come between the Jurassic clays and the Chalk. These are :—

	NAME	CHARACTER	THICKNESS IN FEET
CRETACEOUS	Upper Greensand	Sandy limestone	25 to 0
	Gault	Clay	220
	Lower Greensand	Red and white sands	150 to 0
	Wealden or Shotover Beds	Red sands and clays	50 to 0
JURASSIC	Purbeck Beds	Marls, clays, and thin limestones	15 to 0
	Portland Beds	Limestone and sand	60 to 0

It will be seen from the thicknesses given in this table that the Gault Clay is the only one of all these beds that is never missing. The Gault resembles the Jurassic clays with which it shares the main portion of the sub-soil of the Vale of Aylesbury ; but the frequent presence of hills of sandy and limestone strata, from which the rain washes down material to lighten the clay-soil below, is one of the causes of the great fertility of that Vale.

These hills are nearly all of the same geological character—sandy beds (Lower Greensand or Wealden) at the top ; with Purbeck marls, Portland limestone and then Portland sand below. Such are Brill and Muswell

Hills, Ashendon Hill, Quainton and Oving Hills, the
ridge from Chilton to Long Crendon and that bearing
the Winchendons, and finally the long ridge south of
the Thame from Bierton through Aylesbury and Stone
to Haddenham. Owing to the general dip of the strata,
these hills are highest in the north-west and lowest in

Quarry in Purbeck and Portland beds, Hartwell,
near Aylesbury

(*The Purbeck beds, principally marl, extend down to the level of the
upper platform, the lowest bed, known as "Pendle," being very
clearly distinguished. Below these are the Portland beds, principally limestone*)

the south-east. From the last-named ridge these strata
dip under the Gault and disappear.

On most of these hills the limestone of the Portland
beds either is or has at some time been quarried, as

some layers make a good building-stone while others are burned for lime. The sands of the Wealden and Lower Greensand are also dug in places. This last bed is badly named as it is nowhere of a green colour in Buckinghamshire, though beds of the same age elsewhere show that colour. These sands attain their greatest thickness on the eastern border of the county, at the Brickhills, where they form a conspicuous range of hills, and contain beds of valuable fuller's-earth, which is mined just across the county boundary in Bedfordshire.

The Upper Greensand is a sandy limestone with numerous grains of a dark green substance scattered through it, giving it a greenish look. It is to be seen here and there in the neighbourhood of Princes Risborough.

The Chalk forms the Chiltern Hills and underlies all the country from thence to the south-east. It has the same chemical composition as limestones, but is more powdery in general. Its characters, however, differ greatly in different layers. The uppermost part of the Chalk contains many layers of flint; the lowest layers have none of these and are more marly in character, that is they have clayey matter mixed with the chalk. The fossils of the Chalk, and especially those that are seen when it is examined under the microscope, show that it was deposited at the bottom of a sea which, if not as deep as the centre of the present North Atlantic, was perhaps as far from land. The chalk is quarried everywhere, but in Buckinghamshire only for local use, so the quarries are small.

The Chalk sinks in the south-east under Eocene beds, divided as follows :—

NAME	CHARACTER	THICKNESS IN FEET
London Clay	Clay	70
Reading Beds	Beds of flint-pebbles, clays of various colours, and sands	40 to 80

Chalk Pit, near Penn

(*The Upper Chalk, with bands of flint*)

The London Clay, however, is rarely seen at the surface, as most of its area is covered by Pleistocene river-gravels, and where so covered it is only 20 to 30 feet in thickness.

The strata hitherto described constitute what is

sometimes called the "Solid Geology" of the county, in distinction to the Pleistocene beds which form its "Superficial Geology." These names are not very satisfactory, but they express the broad truth that the Pleistocene beds are so insignificant in bulk that they cannot be shown on anything approaching proportionate scale in a section like that on pp. 36—37. In fact if they could be completely removed from the land the shape of the surface would scarcely be changed. But their absence would make profound differences in the soil, which is greatly diversified by their presence upon the broad areas of clay and chalk.

In the northern part of the county the superficial deposits were in large part laid down during the Glacial Epoch or Great Ice Age, when great glaciers spread downwards from the high grounds of the north of England and invaded part of Buckinghamshire. The most characteristic deposit of this age is the Boulder Clay—a tough clay full of rock-fragments of all kinds and sizes, many brought from a distance. This is found spread over large areas of North Bucks, often concealing the Great Oolite and its associated strata, and covering parts of the great clay-plain. Owing to the variety of materials which it contains, the soil over it is more fertile than that over the stiff Jurassic clays.

Associated with the boulder clay are beds of coarse gravel, deposited by the waters that flowed from the melting ice. These also are scattered over parts of North Bucks, and frequently form the sites of villages, since they are water-bearing strata.

Boulder Clay and Glacial Gravels, Wing

On the Chiltern plateau a great diversity of superficial deposits is found. In places are great sheets of brick-earth (clay mixed with fine sand); in other places clay full of large flints which seem to have come out of the chalk without suffering any wear and tear. Elsewhere there are great expanses of gravel. Many outlying patches of Reading beds are also found, and though these are not strictly "superficial deposits," yet they join with these in greatly diversifying the soil of the Chiltern plateau.

Where the "clay-with-flints" occurs, the soil in the ploughed fields is so full of flints that there hardly seems room for any crops to grow, yet the crops are good. Flints are continually being picked off the fields for road-mending, yet the flints seem as plentiful as ever. Mixed with them are often found well-rounded flint-pebbles, and occasionally blocks of sandstone, both derived from Eocene strata. The sandstone blocks are often of curious rounded shapes, and are known in Wiltshire as "sarsen stones," though that name does not seem to be locally known in Bucks. Such stones may be seen in railway-cuttings, and here and there on a village green. At the Lee, near Wendover, a large one has been mounted on a pedestal. At Walters Ash, near Hughenden, sinking for these blocks forms a regular industry. Blocks of pudding-stone, formed by masses of flint-pebbles being naturally cemented together, are similarly found here and there in South Bucks. They are particularly plentiful near Bradenham. Huge blocks of this kind have been used as corner-foundation-stones at Chesham Church. In the glaciated part of the county, large blocks of foreign rocks out of the boulder clay are sometimes found. One such stands in the middle of Soulbury village, probably brought from Derbyshire by the ice-sheet that spread over the land.

Extensive sheets of gravel are spread both over the high grounds of the Chiltern plateau and the low grounds

Erratic Block from Glacial Drift, Soulbury

(This boulder is nearly three feet high, four feet long, and three feet wide at the base)

bordering the Thames, Thame, Ouse, and Ouzel valleys. These valley-gravels are old deposits of these rivers, laid down when they had not excavated their valleys to their present depth. In them are occasionally found the remains of large animals such as the mammoth, now extinct, and others not now found in Britain ; and with them the primitive flint implements of an early race of Man.

Specimens of the fossils from many of the rock-formations of the county may be seen in the Aylesbury museum.

7. Natural History.

At the time of the Great Ice Age, spoken of in the last chapter, all Britain, except a small part of the South of England, and most of Northern Europe were covered with ice and snow as Greenland is now. As the climate grew gradually warmer, the ice shrank farther to the north, and the plants and animals of Central Europe gradually spread farther north also. There are reasons for believing that, at first, Ireland and Great Britain were not islands, so that plants could spread and animals wander freely into this country from the continent. After a time, however, the destructive action of the waves cut away the connecting land, and Great Britain and Ireland became islands. This took place before the natural process of " stocking " these lands from the continent was complete. Consequently, although our British wild plants and animals are very nearly all of species also found across the English Channel, yet there are many species found there that are unknown in our islands. They did not happen to get across before the separation, and they cannot do so now.

A larger number of species failed to reach Ireland than England, because it was more distant than England.

Plants that grow easily in any soil must have spread more rapidly over the country than those that can only grow well in particular kinds of soil. The whole of the wild plants of our country, collectively called its *flora*, is made up of a number of separate floras of particular soils and situations. The flora of a county can never include so many species as does the flora of the British Isles as a whole, because a county cannot include all the varieties of soil, situation, and climate of the United Kingdom. Some counties have more variety than others. Buckinghamshire, having neither sea-shore nor rocky mountain nor fen within its borders, cannot possibly have the richness of flora of Devonshire or Norfolk. Besides, our county is nearly all cultivated, and the effect of cultivation is to kill off such of the natural plants as cannot thrive in hedgerows or among grass, or grow up quickly and seed early among the crops. Those that can do these things thrive far more in cultivated ground than they did when they had to hold their own against other wild species, and they now constitute the common "weeds."

The principal kinds of soil and situation that may be distinguished in Bucks so far as plants are concerned are (1) water, (2) clay, (3) chalk, (4) sand and (5) gravel, and all of these except the first may be further divided into woodland, grassland, and arable, because the plants that grow well under the shade of trees are not those that grow best in a meadow or among crops.

The water-plants must be distinguished into those found in rivers, as the white and yellow water-lilies (*Nymphaea alba*, *Nuphar luteum*), king-cup or marsh-marigold (*Caltha palustris*), water-avens (*Geum rivale*), sweet flag (*Acorus calamus*), arrow-head (*Sagittaria sagittifolia*), etc.; those found in ponds, as the water-crowfoot (*Ranunculus aquatilis*); and those found in moist, boggy situations, as the sundew (*Drosera rotundifolia*) found at Burnham Beeches, and the snakeshead (*Fritillaria meleagris*) found in the Thame water-meadows and by the Ford Brook at Dinton.

Clay-loving plants are few, and the flora of the clay-plain of North Bucks consists of little more than the species that are common everywhere. Among trees, however, the oak, hawthorn, and holly grow best on a clay-soil, and the woods of North Bucks are mainly oak-woods. In early spring, before the oaks are in leaf, these woods are full of primroses, wood-anemones, and bluebells.

On the chalk we find a much greater number of peculiar plants. On the bare downs and chalky banks various orchids, such as the fly-orchid (*Ophrys muscifera*) and bee-orchid (*Ophrys apifera*), occur, with traveller's joy (*Clematis vitalba*), wild candy-tuft (*Iberis amara*), gentian (*Gentiana amarella*), milk-wort (*Polygala vulgaris*) and many others. A rarity on the downs is the pasque-flower (*Anemone pulsatilla*). Near Ellesborough box-trees grow wild; and throughout South Bucks one may often find the spindle-tree (*Euonymus europaeus*) with its beautiful fruit "which in the winter woodland seems a flower."

The Coral-root

(*Dentaria bulbifera*)

Beech-woods are abundant on the chalk area, and in them are many plants rarely found elsewhere, as the bird's-nest orchid (*Neottia nidus-avis*), butterfly-orchid (*Habenaria bifolia*), helleborine orchid (*Cephalanthera pallens*), Solomon's seal (*Polygonatum multiflorum*), spurge-laurel (*Daphne laureola*), and two plants which though not special to the county are much commoner here than elsewhere in England. These are the coral-root (*Dentaria bulbifera*) and the mezereon (*Daphne mezereum*). The first of these is a slender plant growing to a height of about 18 inches and bearing delicate lilac flowers. It does not often ripen its seed, but multiplies by means of little black bulbils, like those of the tiger-lily. It flowers in spring, and dies down very quickly in the early summer. It is extremely abundant in some of the Chiltern beech-woods. The mezereon is rarely found in the woods now, as it has been so largely removed to ornament the cottage gardens, where it may commonly be seen. It is a winter-flowering shrub, closely allied to the spurge-laurel, but growing larger and more bushy, and shedding its leaves in the autumn. It bears quantities of pretty pink sweet-scented flowers in February, before the leaf-buds open, from which by summer-time bright red poisonous berries are developed.

On a sandy soil fir-trees grow well. If there are no trees, there are heathy commons with ling and gorse and broom. The chief area of this kind of land is on the borders of Bedfordshire, about Wavendon, but there are many smaller areas on the Chiltern plateau. Gravel-soil has a similar flora to the last, but with greater abundance of the common plants which do not thrive at all on a dry sandy soil.

To those who live among the wild flowers, as well as to those who may know them only from short visits, some words of advice may be addressed. A rare flower should never be rooted up, and should only be picked if it is known that (as with many of the orchids) they are propagated less by seed than by other methods. Those who wish to learn all they can about wild plants will make a point, whenever they see a flower unknown to them, of noting carefully the soil and situation where it is growing, the time of flowering, the forms of the stem and leaves, and the way the flowers are set on the stem, as well as the characters of the flower itself. In this way they will not only find it easier to determine the name of the flower, but will be able to look out for it again and recognise it in new places.

The trees of Bucks have very largely been planted by man. The elm rarely or never seeds in England, so that it must have been introduced by man, and it is still only found near villages or farms or by roadsides, though its habit of sprouting from the root enables it to spread along hedgerows. In many places, especially in North Bucks, avenues of elms as at Thornton, or limes and horse-chestnuts as at Wavendon, have been planted, more rarely of beeches as in the northern part of Stowe avenue.

The famous Burnham beeches owe their peculiar form to having been pollarded at some past time. The yew

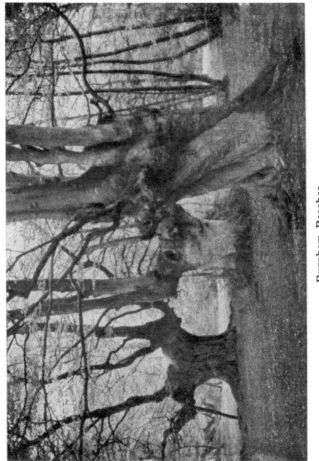

Burnham Beeches

has been planted in churchyards from time immemorial, and there are ancient yews of immense girth in the churchyards at Ibstone, Hedsor, and Langley Marish.

Wych elms (*Ulmus montana*) are found in the Thames valley and about Brill, and the hornbeam (*Carpinus betulus*) is a common tree everywhere. Among famous oaks

Waller's Oak, Coleshill

must be mentioned Waller's Oak at Coleshill, and Cowper's Oak in Yardley Chase.

Its beech-woods are the chief glory of South Bucks, for they are beautiful at all seasons. In winter, the bare branches massed together show in distant view a dark purple colour, which varies with the varying light and shade. On nearer approach, the gloom of the woods is broken by the pale grey-green of the columnar trunks, and the green of the undergrowth—dull green holly and more vivid bramble—while beneath all is the rich brown carpet of last year's leaves, changing through chocolate to black as the season passes. In spring-time, no tree is more individual than the beech in its time of budding. As the buds swell, the distant purple becomes spotted with grey, changing quickly now here and now there to light brown and then to the most delicate of pale greens, while in places burst out brilliant white patches of wild-cherry blossom. It is now that the interior of the woods is most exquisite in its beauty. A fresh brown carpet has been spread— a carpet of fallen bud-scales, the undergrowth shows a more plentiful green, while primroses and other flowers are scattered about. Above, the sunlight shines through the translucent pale green of the newly-spread leaves, each still bearing the silky fringe along its margin. As summer comes, the fringe falls off, the leaf becomes thicker and darker and less translucent, the whole wood is more sombre, the flowers are fewer. Still it is only the recollection of May-tide that can arouse any dissatisfaction with the beauty of the wood.

Before summer is gone, a yellowing of the leaves already begins in individual trees, but from mid-October to early November is the time when the greatest glory of colour is to be seen. The tints of the foliage range from yellow to golden brown, and here and there rich crimson of wild cherry or pale yellow of elm adds to the variety. The aspect changes with change of light and shade, and the passing days make new masses of leaves change colour while the earlier victims fall to make a fresh carpet once again. Meanwhile the soil below, as if in emulation of the branches above, sends up singly and in bunches fungi of most varied hue—scarlet and yellow, orange and purple, white and brown. Autumn is the most gaily-coloured season in the beech-woods. Gradually the leaves all fall, except from the short branches near the ground, covering up the black decaying fungi, and the winter aspect of the woodland is resumed once more.

There are no special features to be noticed in the mammals of our county, and the same may be said of the birds, which include the usual species of Southern England. In South Bucks, nightingales are very plentiful, and the woods and waste spaces afford protection for the night-jar, green woodpecker, goldfinch, magpie, jay, and many commoner birds, in spite of the large amount of game-preserving which tends to exterminate some of these birds.

The inland situation and the elevation of a considerable portion render our county less rich both in species and individuals of its insect fauna than are the eastern and southern counties, which lie nearer to the coast and are therefore subject to constant immigration from the continent as well as having many species indigenous to a littoral area. Yet, since in a large part of the county the chalk formation is at the surface with its special flora thriving only on soil rich in calcareous matter, a number of species are found whose earlier stages can only subsist on these local plants. To this category belong two of the most beautiful of our native butterflies, the Chalk-hill

Blue (*Agriades coridon*) and the Adonis Blue (*A. thetis*), the former of which occurs on most of the chalk escarpments of the county. Wherever the woolly mullein plant (*Verbascum thapsus*) flaunts its brilliant yellow spikes, there may generally be found the conspicuously beautiful pale blue larvae with yellow and black spots of the Shark moth (*Cucullia verbasci*). This is not wholly confined to the chalk however, for the abundance of fig-wort (*Scrophularia aquatica* and *S. nodosum*) growing in the clay areas in proximity to the chalk, gives ample nourishment to the same species. On the sunny slopes of the chalk, towards the end of July, one may in many places meet with the Six-spotted Burnet moth (*Anthrocera filipendulae*) flying in its characteristic slow heavy way from one flower-head to another, appearing to be conscious of security from attack, which the acrid fluids of its body ensure, and its brilliant glittering bronzy-black and rich purple advertise so well. The yellow-flowered rock-rose (*Helianthemum vulgare*), exclusively a chalk plant, is the food of the young stage of the rare Green Forester moth (*Adscita geryon*), said to be found near Tring, while the woods of mingled beech and wych elm in the neighbourhood of Amersham produce two other species (*Asthena bloomeri* and *Abraxas grossulariata*), but rarely found south of Yorkshire.

The fact that most of the county is closely cultivated and but little of it at any time lying waste, while there are but few commons or open forest lands, limits the number of species to a considerable extent, since insects are wholly dependent for their existence on the presence of the particular plants necessary for the food of their larval stage, and plants can only flourish on soil congenial to them, where their struggle is more or less equalised, and when they can be left undisturbed by the customs of cultivation.

8. Climate.

By the "climate" of any area is meant a generalised statement of all those varying natural conditions which make up the daily "weather." The chief of these are sunshine, wind, rain, and temperature. Three main circumstances combine to determine the climate of any place, (1) its latitude, (2) its position in relation to the great areas of high and low atmospheric pressure on the earth's surface, and (3) its position in relation to land and sea.

The British Isles lie in the temperate zone—Buckinghamshire lying between 51° 26′ and 52° 12′ N. latitude. Consequently the sun in our county stands about 62° above the horizon at noon on the longest day, and 15° at noon on the shortest day. We are thus free from the extreme conditions of temperature of the Torrid and Arctic zones.

The British Isles lie between a very constant area of low pressure covering Iceland and Greenland and a fairly constant high-pressure area covering Central Europe. Consequently the most frequent winds are those which circle in a counter-clockwise direction round the former area, blowing from the south-west in their course across Britain. But frequently small low-pressure centres, or *cyclones* as they are called, travel from west to east across the country, bringing a series of wind and weather changes in their course; while at other times the European high-pressure area extends over our islands, bringing clear cloudless skies with little wind, with great heat and drought in summer and hard frost in winter (anticyclonic weather).

Thirdly, the British Isles have the Atlantic Ocean to the west and south-west, that is to windward in general. The prevailing winds are therefore winds loaded with water-vapour, in other words rain- and cloud-bearing winds. It is only when the usual distribution of atmospheric pressure is temporarily reversed that we get the dry east winds from the continent, or the cold, snow-bringing north winds from the Arctic Ocean.

Another result of the nearness of the Ocean is that our temperature changes less from summer to winter than does that of inland areas. This is for two reasons —firstly because water needs more heat to raise its temperature a given amount than other substances, and conversely gives out more heat as its temperature falls; and secondly because the circulation of water distributes the heat received in lower latitudes to colder regions.

But there is a more powerful equalizer of temperature in the cloud and rain of our south-westerly winds, which in summer keep the air cool by screening off the sunshine, and in winter keep it warm with the heat which they have brought in a "latent" form from warmer latitudes. Thus it is that the ports on the western side of each of the continents are ice-free up to far higher latitudes than those on their eastern shores.

In summer the *isotherms*, or lines of equal average temperature, on a map of the British Isles run generally east and west, the temperature falling pretty steadily as the latitude increases, because sun-warmth is the main factor in summer temperature. In winter, on the other hand, it is rain-warmth which is the chief factor, and the isotherms run roughly north and south—both temperature and rainfall decreasing steadily as we go eastwards.

The geographical distribution of rainfall during any one storm may be of either of two kinds. When the rain falls in connection with a small low-pressure system (cyclone), its distribution depends entirely on the distribution of pressure; so that the rain from a series of such systems will in the course of a year or more tend to be equally distributed over the whole country. When however rain falls in connection with the general south-westerly wind-drift, the greatest fall will be on the high grounds on the western side of our islands, and there will be a fairly steady decrease of rain eastwards. This eastward diminution is well shown on the map opposite which is based upon averages taken over a long period of years.

Buckinghamshire is an inland county lying towards the eastern side of England. Its rainfall is therefore much less than that of the hilly western counties facing the Atlantic winds. As the map shows, it has a rainfall nearly everywhere of between 25 and 30 inches a year; but the rainiest region, the Chiltern crest between Princes Risborough and Tring, seems to have over 30 inches, while the Slough neighbourhood and a small area near Wolverton may have less than 25 inches. The absence

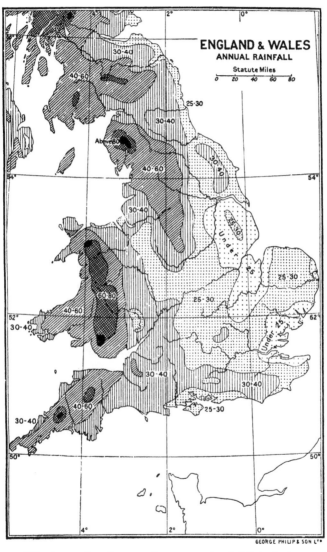

ENGLAND & WALES
ANNUAL RAINFALL
Statute Miles
0   20   40   60   80

30-40
40-60
25-30
30-40
Above 80
40-60
30-40
30-40
30-40
Under 25
25-30
25-30
Under 25
60-80
40-60
30-40
30-40
40-60
30-40
30-40
25-30
GEORGE PHILIP & SON LTD

(The figures give the approximate annual rainfall in inches)

of a sufficient number of rainfall observers in the county makes more precise statements impossible.

At Upton, Slough, where careful observations have been made for many years, the driest year was 1901, when only 16·868 inches of rain were recorded, and the wettest was 1903 with more than twice that amount, 35·860 inches. The rainiest twelve months, however, were October 1878 to September 1879, with exactly 40 inches. The driest month recorded was February 1891, with 0·022 inch; the wettest, October 1903, with 6·440 inches. The wettest day was the 17th July 1890, when 2¾ inches fell. The largest number of wet days in a year was 201 in 1909; the least, 129 in 1893.

At Berkhampstead, which though in Hertfordshire is almost surrounded by Bucks, the mean yearly rainfall for the last 54 years has been 28·56 inches, with a maximum of 38·97 inches in 1872 and a minimum of 19·12 inches in 1898. The driest month was again February 1891 (0·05 inch), and the wettest, October 1891 (8·04 inches).

The yearly average of bright sunshine at Slough is 1464 hours; at Berkhampstead, 1470¼ hours. The sunniest year at Slough was 1899 (1754¾ hours), at Berkhampstead, 1893 (1761½ hours). The brightest month at Slough was July 1911 (329⅓ hours), and the dullest, December 1903 (15·46 hours); at Berkhampstead, May 1909 (290¾ hours) and December 1890 (3¾ hours). The greatest quantity of sunshine recorded on one day, at Slough, was 14½ hours, on June 11, 1898, when the sun was above the horizon for about 16½ hours.

The mean annual temperature of the air in shade at Berkhampstead is 48·3° Fahr., with a mean summer temperature of 59·7° and a mean winter temperature of 37·7°. The mean annual temperature for the whole of England is 48°, so our county is slightly warmer than the average. These seem low figures, but it must be remembered that they include night as well as day temperatures. The highest recorded shade-temperature at Slough was 97° on July 15, 1881, and this was all but reached again on August 9, 1911; the lowest temperature, 0° in February 1895.

9. The People of Buckinghamshire.

Before written history begins to tell us anything of England, we know that the British Isles had long been inhabited by various races of men. They have left behind them remains of their handiwork from which we can form some idea of their mode of life.

The oldest race of which we have any clear evidence in Buckinghamshire dates back to the later part of the Pleistocene period when the climate was colder and rainier than it is now, and so far back in time that the rivers had not worn down their valleys to the present level, but flowed sometimes 50 feet or more above it. In the gravels left by these rivers we find stone implements, chiefly made of flint, roughly chipped into shapes useful to man. These men have therefore been called the Palaeolithic (i.e. Ancient Stone) race.

For how many centuries the Palaeolithic race occupied

the country we cannot guess, but it must have been a very long time. The later races all lived in a Buckinghamshire that differed little from that of to-day. First of these was the Neolithic race, who still used flint for their tools, but usually worked them smooth and polished them. They perhaps first made some of our ancient roads and track-ways still in use. Then race after race entered Britain from the continent, driving the previous inhabitants before them to the west and north. First came the Goidelic or Gaelic Celts, who introduced bronze tools and weapons, and whose descendants still survive in the Scottish Highlands. Then came the Brythonic Celts, who used iron; their descendants still live in Wales, mixed with older races. They were still the dominant race in south-east England when our written history begins with the invasion of Julius Caesar.

During the four centuries in which Britain formed part of the Roman Empire, soldiers and traders of many races must have left descendants here to mix with the native races; but when the Roman legions had been withdrawn about 420 A.D. a fresh invasion took place— that of the Saxons and Angles. Opinions differ as to how far the Brythonic people or Ancient Britons were driven out of Buckinghamshire, how far they were killed off, how far they survived in the wooded areas, and how far they remained as the slaves of their Saxon conquerors.

Of the people who now live in Buckinghamshire, some have come there from elsewhere, some are the children or grandchildren of men or women who came from elsewhere, while some belong to families that have

lived in the county for many centuries. Whatever may
have been the races that peopled the county during the
many centuries when there was comparatively little
migration from place to place (say from the seventh
century to the seventeenth), the present population is
of very mixed descent. This mixed character is greatest
in the industrial towns, such as Wycombe and Wolverton;
and least in the agricultural villages that are farthest
removed from town influence. It seems probable that
in the Vale of Aylesbury, at least, there is still a large
part of the population that is descended from the Saxon
conquerors who settled in the district towards the end of
the sixth century. Such was the impression made upon
the skilled observer of race-characters, the late Dr Beddoe,
by the people at Aylesbury market. In the Chiltern
district, the impression produced on a casual observer is
that there is no uniformity of type—very dark and very
fair persons are both common; and statistics show that
there is much more of the dark element in the population
than there is in the Vale. This may be due to the fact
that the Saxons settled more gradually and in a more
peaceable manner in the Chilterns than in the Vale, and
did not drive away or exterminate the British. On the
other hand it may be that in quite recent times there have
been more persons coming from distant parts of the kingdom
to settle in the Chilterns than in the Vale.

So far as historical evidence goes, we must give pre-
eminence in Buckinghamshire to the Saxon race. Some
traces of their predecessors the British or Brythons there
probably are. The Danes never seem to have settled

here to any appreciable extent. The Normans formed
an aristocracy only, and there has been no later immigration of any particular race or people in a body, though
from the seventeenth century there has been some influx
of people of the most varied origin into South Bucks from
London.

The earliest date at which we can form some idea of
the numbers and distribution of the population in Bucks
is the end of the eleventh century, when Domesday
Book was compiled. This records the numbers of all
the cultivators of the soil in every township, but it is not
a census, and we must guess at the numbers of women
and children. If we assume that every man mentioned
in Domesday had, on the average, a wife and four
children or aged parents dependent on him, we shall find
the total population of the county to have been about
30,000. To-day it is over 200,000.

In the eleventh century the most thickly populated
district was the valley of the Ouse, where there were
probably over 50 persons to the square mile. In the rest
of North Bucks there were rather less than 50, and in
the Chiltern Hundreds between 20 and 30 only. If we
compare these figures with what is shown on p. 72, we
see that there are still parts of North Bucks that are very
possibly no more thickly populated than they were eight
centuries ago, but they were among the most populous
places in the county then, and are among the least so
now. The balance of population has now shifted to the
Chiltern district, and here and there all over the county
towns have grown up, any two or three of which contain

as many inhabitants as did the whole county at the Norman Conquest.

It is only during the last hundred years that accurate censuses of the population have been taken. By means of these we can trace at ten-year intervals the gradual change in numbers and their distribution about the county. From 1801 to 1901 the population increased from 107,900 to 195,900. The general history of our villages is very well illustrated by the upper curve on p. 70 (Whitchurch). We see a steady increase of numbers

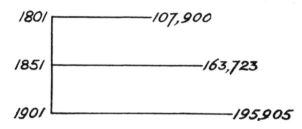

1801 ——————————*107,900*

1851 ——————————*163,723*

1901 ——————————*195,905*

Growth of Population of Buckinghamshire in the
Nineteenth Century

during the first half of the nineteenth century, due to the excess of births over deaths and to the absence of any large amount of emigration. The second half of the curve tells the tale of the "rural exodus," which is associated with the enormous growth of our industrial towns.

The change from village to town is strikingly illustrated by the diagram of the population of Wolverton through the nineteenth century. Up to 1831 the population and the rate of its change is practically the same as that of its neighbour, Haversham. Then comes the London and Birmingham Railway, with its engine and carriage-building works, and after that Wolverton shoots up into a town while Haversham remains a small village.

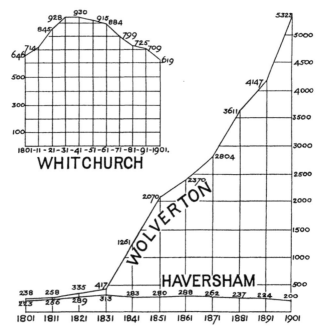

Nineteenth Century changes in Village and Town

Buckinghamshire contains several almost uninhabited parishes. At the time of Domesday Book these were all occupied by agricultural villages, but the turning of arable land into sheep-grazing land in the fifteenth century depopulated them, and they have never recovered. The most striking case is Creslow, with a population of 5, but Tattenhoe (16), Grove (19), Horsenden (35), Pitchcott (40), Stantonbury (41), Aston Sandford (46), Foscott (46), Pounden (46), Ilmer (51), Fleet Marston (53), Hogshaw (56), and Quarrendon (65), are also remarkable. In four of these not only has the village disappeared, but the church is in ruins (Quarrendon) or has disappeared altogether (Creslow, Hogshaw, Pounden).

The population of Bucks in 1901, the date of the last census of which the returns are complete, was, in round numbers, 196,000, of which 95,000 were males and 101,000 females. Of the total, 43,000 were under 10 years of age, and of the remainder, 71,000 were classed as unoccupied and 82,000 as occupied. This seems an alarming number of idlers, until we realise that the " unoccupied " include all school-children and all students over 10 years of age, as well as all girls and women engaged in unpaid domestic work.

Sixteen thousand persons, or nearly one-fifth of all the " occupied," were engaged in agriculture ; 15,000 in domestic service (paid) ; 7000 in transport (by railway, road, canal, and river) ; and 6600 in all kinds of building and constructive work.

These are occupations common to all counties, in proportions which do not vary greatly from one to another. Now we come to some more special Buckinghamshire occupations. Furniture-making and woodwork occupied 5300 persons ; railway-carriage and coach-building, 2700 ; boot and shoe-making, 2300 ; brick,

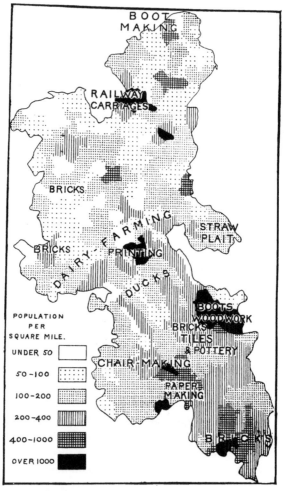

Population and Industries of Buckinghamshire

tile, and pottery-making, and printing, 900 each; tanning and lace-making, 800 each.

In 1911, the total population had increased to 219,583, of which 107,343 were males and 112,240 females.

10. Place-names of Buckinghamshire.

Names have been given to places at all times since men have dwelt in them, and the names to be found on the map of Buckinghamshire are a mixture of all dates. When we find a farmhouse called Inkerman, or a cluster of cottages called Gibraltar, we may make a good guess as to the dates when those places were built; but many of the names are so ancient that it is not easy to be sure when they were given. There are about 230 names of parishes and hamlets in the county, and of these nearly 200 are found recorded in Domesday Book, more than 800 years ago. The names have changed a little in that time, but they are easily recognised. Nor can we doubt that many of them were very old names even in the time of the Conqueror; for occasionally we find mention of some of them in still older documents.

Perhaps the oldest village-name in Bucks is that of Kimble—the name of two villages at the foot of the Chilterns, distinguished as Great and Little. Tradition connects this district with Cunobelinus or Cymbeline, a British king who ruled in eastern England about the beginning of the Christian era and whose name gives the title to one of Shakespeare's plays; and when we find a document dated A.D. 903, calling these villages *Cunobelingas* and Domesday Book naming them *Chenebelle*, we can scarcely doubt that tradition is right.

The names of the chief rivers may be even older, for the names Ouse and Wye are modifications of the Celtic word for water, and must have been given to those rivers by a Celtic-speaking race before the English came to the land ; while the name of the Thames is given as *Tamesis* by Julius Caesar (B.C. 55). There are a few other names that may just possibly have been given in the days when a Celtic language was spoken, and retained by the English-speaking people who conquered the land, but they are doubtful. Wendover, for instance, sounds very like a Welsh name, but we cannot be sure of it.

The great majority of the village-names were given by English-speaking people, and are therefore not older than the last quarter of the sixth century, when the Saxons seem to have first settled in this district. When a Saxon warrior with his family and his slaves settled down to cultivate an area of land, he called the settlement his *ham* (home), and it often came to be known by his name—thus the home of Agmond was called *Agmondes-ham*, which in time became shortened to Amersham. In other cases he surrounded his house and yard by a hedge or *tun*, and in time the word *tun* came to mean a *town* or village ; and there are more village-names ending in *ton* than in anything else in Buckinghamshire. The maps on p. 75 show that *tons* are scarce in the Chiltern district, while *hams* are chiefly found there. This seems to suggest that the first English settlements in this district

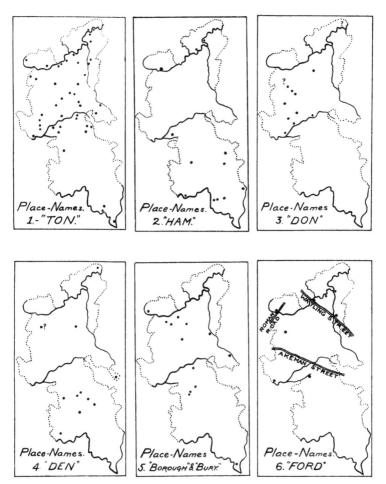

Place-Names.
1.-"TON."

Place-Names.
2."HAM."

Place-Names
3."DON"

Place-Names.
4."DEN"

Place-Names
5."BOROUGH"&"BURY."

Place-Names.
6."FORD"

Terminations of Place-Names in Buckinghamshire

took place at a different time and under different conditions from the settlement of the northern plain.

The terminations *don* and *den* show a similar oppositeness of distribution. *Den* or *dene* means a wood, while *don* is a down or bare hill or sometimes an artificial mound. In the plain the settlers chose rising ground to live upon; in the dry Chilterns they more usually settled in the wooded valleys. It seems reasonable to suppose that the villages whose names end in *don* and *den* are younger than those ending in *ton* and *ham*, because if the downs and denes had not already had names given them by earlier English settlers, the new settlements might have been named *ham* or *ton*.

Another set of English village-names are those ending in *borough* and *bury*, meaning a dwelling fortified by a moat or stockade, the first part of the name being usually the personal name of the man who built the fortification. The map again shows these names to be almost confined to the plain.

A few names end in *ey*. This termination is common in the Fen district, and along the valleys of large rivers with a broad alluvial margin. It means literally an island, and was applied to settlements near the water but just safely out of reach of floods. The few places with this termination in Bucks are, of course, along the Ouse, the Thame, or the Thames. There is however another termination which must not be confused with it—*ley*, a lea or open ground. This termination is confined to the northern plain and the Thames valley.

The name *ford* explains itself. It occurs as a

name-ending chiefly along the Ouse, the Thames not being fordable along the Bucks border. Twyford, the only "ford" that the map does not show near a river, is near a large brook tributary to the Ouse. Where the name *Stratford* occurs we shall always find a Roman road or *Street* crossing a river.

Other terminations that are found here and there are *cott* or *cote* (a cottage), as in Edgcott, Foscott, Pollicott; *low* (a tumulus under which some great chieftain was buried), as in Bledlow, which is called in documents of the eleventh century Bleddanhlaew, that is, Bledda's tumulus; *wick* (a little village), as in Tetchwick and Tingewick; *well* (a spring of water), as in Hartwell and Bradwell.

There are no village-names in this county ending in *by* or *kirk*, terminations that are plentiful in the Eastern counties and are of Danish origin; and of the other common Danish ending, *thorpe*, there are but three examples—Eythrope (in Waddesdon parish), Helsthorpe (in Wingrave parish), and Tythrope (in Kingsey parish). We learn from this that although the Danes harried Buckinghamshire on several occasions, and although the county was reckoned as one of the shires of the Danelagh, the Danes scarcely settled here at all.

It often happens that two villages have the same name, and to distinguish them either a surname is added or some descriptive word prefixed. From Domesday Book we learn that these methods had scarcely begun to come into use at the Conquest, and the surnames are generally those of the great families who owned the villages in the thirteenth century.

Thus we find Stoke Mandeville, Stoke Poges, Stoke Hammond, and Stoke Goldington ; Aston Clinton, Aston Sandford, and Aston Abbots (the last name referring to the Abbot of St Albans) ; Drayton Beauchamp and Drayton Parslow. Sometimes the owners' names come in front, as in Monks' and Prince's Risborough. Grendon, Wotton, and Weston Underwood get their surname from their nearness to one of the great forests. Stony, Fenny, and Water Stratford ; Steeple, East, Middle, and Botolph Claydon ; Fleet and North Marston show other ways of distinguishing. Great and Little distinguish two out of the three Brickhills, and the two Kimbles, Hampdens, Missendens, Marlows, and Linfords. The two Chalfonts bear the names of the patron saints of their churches— St Giles and St Peter.

In one case two adjacent villages have dropped their original name and kept only their surnames—Iselhampstead Chenies and Iselhampstead Latimer being now known as Chenies and Latimer simply.

The spread of a general knowledge of reading is tending to destroy the customary methods of sounding local names, where these are different from the sounds suggested by the spelling ; but we can still hear Chalfont pronounced *Charfunt*, Coleshill, *Cōsul*; and Chesham— as it should be—Ches-ham (*Chessam*).

11. Agriculture.

In the year 1910, nearly 396,000 acres were under cultivation out of the total 479,359 acres in the county of Buckingham; or in other words over 83 per cent. of its area. This is above the average of England, which as a whole has only about 76 per cent. of its area cultivated, and in fact there are only about twelve counties which have a higher proportion cultivated.

This 83 per cent. may be divided into two parts— 54 per cent. of the whole area is permanent grass, while 29 per cent. is arable land—that is, land that is regularly ploughed. Every year, for many years, more land has been laid down in grass, and every year less land is ploughed—and the increase in the former is never quite equal to the decrease in the latter. Consequently the total cultivated area is steadily diminishing, although it is still so high. This is because, every year, more land is wanted for houses, roads, railways and so forth ; not because any of it is being allowed to go to waste. In Bucks also there is a steady increase in the area of plantations.

Buckinghamshire thus has more than half its area in permanent grass ; while England as a whole has only 45 per cent., and such a county as Suffolk has only 20 per cent. Buckinghamshire thus takes its place among the grazing rather than the ploughing counties ; but if we could get separate statistics for the northern and southern divisions of the county, we should find that while North Bucks showed a still higher percentage of grass-land, South Bucks would stand higher in its proportion of arable land. It would not however approach such a county as Cambridgeshire, which has 67 per cent. of its area under the plough.

The great clay-plain of North Bucks is thus a great area of grass-land, on which cattle are grazed for milk and meat. The Vale of Aylesbury especially is famous for its dairy-farming. Apart from this, there is not much about its agriculture that particularly distinguishes Buckinghamshire from other counties. It is in fact a fairly average county, and is not famous for any particular crop, or any other product except the ducks of the Vale of Aylesbury. Even where a local breed has become famous, as in the case of the Oxford Down sheep of Upper Winchendon, it must be credited to the individual breeder rather than to the county.

Of the arable land, nearly a quarter is sown with wheat—a proportion which is much above the average of England, and not far below that of the greatest wheat-growing counties[1]. Less is devoted to barley than the average of England; and to oats, rather more than the average[2]. Much less than the average is devoted to turnips and potatoes. Next to wheat and oats, clover is the chief crop : these three take up over 60 per cent. of the arable land. Among minor crops, beans and peas,

[1] The exact proportions are—Huntingdonshire (highest percentage), 26 per cent.; Bucks, 23; all England, 15.

[2] Barley, in all England, takes 13 per cent. of the arable land; oats, 18 per cent. In Bucks the percentages are—barley, 10; oats, 19.

grown for fodder, take nearly double their average share of the arable: these, especially beans, being familiar crops in North Bucks. Of turnips and potatoes there are less than the average, of mangold and cabbage just about the average, of vetches rather more than the average, and of lucerne (grown for pasture and hay) more than double the average.

In fruit growing, Buckinghamshire is also an average county. In England, taken as a whole, there are for every thousand acres $7\frac{1}{2}$ acres of orchard; Buckinghamshire has rather more than $7\frac{3}{4}$ acres in every thousand. This area is made up of $2\frac{1}{4}$ acres of apple-orchards, $1\frac{3}{4}$ cherry, $1\frac{1}{4}$ plum, $\frac{1}{2}$ gooseberries and currants, and the rest mixed. This is far less than the average for apples, gooseberries, and currants, and far more than that for cherries and plums.

A crop of much importance in the Chess valley, though not strictly an agricultural product, is watercress; for the growth of which the slow-flowing river and artificial water-ways cut in the adjacent levels are well adapted.

As might be expected from the large amount of grazing land, Buckinghamshire has more than the average proportion of cattle—173 to every thousand acres, as against 156 for England. It is a long way from being a great cattle-raising county like Cheshire, which has 281 per thousand acres, but it is in much stronger contrast with Suffolk, which has only 84. Of sheep, Buckinghamshire has rather less than the average—380 to every thousand acres, instead of the 458 for all England. Of pigs and of horses used in agriculture it has just about the average. Of poultry there are no statistics, but there are few large poultry-farms. The Vale of Aylesbury, a century ago, was noted for its ducks, the clay soil affording abundant ponds for them. At that time every cottager raised his brood of ducks, which in many cases shared his cottage with him and his family. We are more civilised now, but duck-farming is still an industry of some importance, especially at Monks Risborough and Weston Turville.

Of the future of Buckinghamshire agriculture it is not safe to speak, though we may safely prophesy great changes. The opening up of parts of the county by new railways, the growth of London and other towns with their demand for milk, vegetables, etc., and the diminishing demand for hay owing to the displacement of horses by electric and petrol traction, are all beginning to exert their economic influence. Another important change is likely to result from the steady increase in the number of small holdings, with the more intensive cultivation that they bring. At Michaelmas, 1910, there were over a hundred small holdings under the Bucks County Council, with an average of 20 acres each, to which must be added many that were otherwise obtained. There were also 13 properly constituted Agricultural Cooperative Societies —very necessary aids to the success of small holdings.

Duck Farm, Kimble

(Chiltern Hills in the distance)

12. Industries and Manufactures.

It has already been explained why Buckinghamshire is not an industrial county, although it has a number of small industries. It will be well, however, to consider carefully what is meant by industry and manufacture. No community can live without working, nor can they work without tools; so that, in one sense, the work of agriculture is industry, and the making of tools by the village blacksmith is manufacture. In England, a thousand years ago, the great bulk of the people were engaged in agriculture, but carried on such handicrafts as the weaving of cloth and the making of simple tools, especially during the winter. Every village produced for itself very nearly all that it wanted; but even then there were some things, particularly metals, which could not be obtained everywhere, and other things which could be raised more plentifully in one place than another. To buy and sell such things markets and fairs grew up; and though fairs are nearly done away with nowadays, markets are as useful as ever. As commerce extended, it became more and more possible for people in one place to devote themselves to making one kind of thing well, and selling that to buy all the various things they wanted, instead of making all things for themselves. In this way there came to be many small local industries, owing their existence either to the soil being specially suitable for the growth of the raw material, or to the presence of water-power. In later times, immigrants from abroad settled in some parts of the country and brought some new industries with them; but Buckinghamshire seems to have gained very little in this way.   In other cases, some local man of exceptional ability or novelty in ideas started some new industry in his own district.

The industrial revolution of the end of the eighteenth century, when steam-power and machinery on a large scale were introduced, caused the transfer of many industries to the coalfields; but in some cases local industries elsewhere were strongly enough rooted to continue even at the cost of bringing the coal to them instead of migrating to where it is found.

The chief local industry of Bucks is that of furniture- and particularly chair-making, which in 1901 occupied some 5300 persons (just one-third of the numbers engaged in agriculture).   It prevails in South Bucks, and is dependent on the Chiltern beech-woods.   It still shows various stages in industrial development.   Near Wendover and Bradenham, for instance, men still work in rough sheds in the woods, cutting up the tree-trunks and turning chair-legs on the primitive pole-lathe.   In this the rotating movement is imparted by a rope twisted round the object to be turned and actuated by a foot-treadle, so that the direction of rotation is reversed at each movement of the treadle, and the shaping tool is idle half the time : the turners claim that they can use their tools more effectively on this lathe than on the continuous wheel-lathe.   In the small villages around High Wycombe there are still many little workshops where the parts of chairs are made by hand and by the

Roadside Chair-factory, Whielden Gate, Amersham

pole-lathe, fitted together, and stained; and huge cartloads of wooden chairs may be seen on the main roads between this district and London. In the same villages, the women and girls may be seen making the seats of cane chairs. But in High Wycombe the industry has developed on a large scale. Here are large factories, which get their wood not only from the woods near at hand but even from abroad; and make, not chairs only, but furniture of all kinds, some of which is exported. The chairs used at the Coronation of King Edward VII were made in Wycombe.

Beechwood is suitable for other articles than chairs, such as those used in dairy-work, the blocks of brushes and the wooden pails and spades used by children at the sea-side. Chesham is the chief centre for the manufacture of these articles. Tennis-rackets are made at various villages.

The straw grown on the flinty soil of the Chiltern plateau is reputed to be the best in England for weaving, and during a large portion of the last century the splitting and plaiting of straw was an industry practised in nearly every cottage in the Chiltern areas of Bedfordshire and the adjoining part of Bucks, as well as in the Vale of Aylesbury about Stewkley and Hoggeston, and at Amersham. Although it has lost its importance now from the extensive importation of Oriental straw-plait, it still persists on the Bedfordshire borders and as far west as Weston Turville, whence straw-plait is sent to the straw-hat factories at Luton.

The suitability of the clay-plain of North Bucks for

Pole-Lathe in a Village Workshop

(The pole is 15 feet long, but only the free end can be seen, in the upper part of the picture. It plays the part of a spring, the cord which is attached to its end being twisted round the chair-leg that is being turned, and actuated by the treadle)

pasturing cattle and the abundance of oaks in its woods to furnish tan led to the development of the tanning industry there, but it has dwindled away now.   Northamptonshire, which is still more famous as a grazing county, has become one of the great counties for leather-indus-

A Load of Chairs
(*On the way to London*)

tries, particularly boot and shoemaking; and this industry has lately spread across the county-boundary into North Bucks, where at Olney boot-making factories have been established.   The same industry has also been established for years at Chesham, in the south.

Buckinghamshire is not a county with much water-power, but what there is has been utilised, not only for flour-mills, which may be found on all the larger streams, and saw-mills, which are of importance at Wycombe and Chesham, but also for paper-mills, the chief of which are in the lower part of the Wye valley, at Wycombe and Wooburn Green, and in the Colne valley at Horton and Wraysbury. The nearness of London, where so much paper is used, explains the presence of this industry for more than 300 years here. There are also two paper-mills on the Lyde brook, a small tributary of the Thame, at Bledlow.

The paper made in South Bucks is not all sent at once to London: some of it makes a journey to Aylesbury to be used in the printing of books. Some of the finest modern examples of the art of printing, including colour-printing, and of book-binding, are turned out at Aylesbury; so that the county can boast that what enters it as esparto-grass leaves it as a printed book. The establishment of printing-works at Aylesbury (apart from small presses for local printing) only dates from 1867, and is a result of the tendency of London manufactures to move out when they can to country-towns not too far from London. Another paper-industry is found at Wolverton, where the envelopes and stationery for the London and North-Western Railway are manufactured.

Another effect on Buckinghamshire of the nearness of London is that important main roads, railways, and canals pass through it, as do also the two important rivers, the Thames and Ouse. Consequently a considerable part of the population is concerned with ministering to the needs of travellers.   Thus the building of railway carriages, motor-cars, and horse-driven vehicles occupies nearly 3000 persons, and boat-building is also an important industry.

Wolverton is the great railway-coach building town.

Printing Books at Aylesbury

It happens to lie about half-way between London and Birmingham on the original railway between these two places, opened in 1838 and now known as the London and North-Western Railway.   For this reason, in the early days of the railway, all trains stopped there for the passengers to obtain refreshments while engines were

changed and the carriages inspected. It was therefore
a most convenient place for engine-building and repairing
shops. In later years the London and Birmingham Railway Company was amalgamated with the Grand Junction
Railway Company (from Birmingham to Liverpool and
Manchester) which had similar works at its half-way
station, Crewe. A division of labour was then made : all
engines were made at Crewe, and only carriages and
minor articles at Wolverton (p. 150).

There are small carriage and cart-building works in
many places, but the only large ones are at Newport
Pagnell, where are also motor-car works—Newport being
on the main road from London to Northampton and
Leicester.

Boat-building is chiefly carried on at Eton on the
Thames, most of the boats being pleasure rowing-boats,
not only for local use, but for export to all parts of
England and even to foreign countries. More surprising
is it to find that such an inland place as Stony Stratford
takes its place among ship-building towns—making steam
yachts, tugs, and launches for all parts of the world.
These boats have to be hauled on a steam-trolly to the
Grand Junction Canal, but once on the water can steam
to any port in the world.

One of the most famous industries of the county is
that of lace-making. This is believed to have been
introduced by Queen Katharine of Aragon, and to have
been carried on by immigrants from Flanders. In the
seventeenth and eighteenth centuries it was the most
important local industry : almost every cottage in the

county made lace, and lace-making was taught to little
girls in school just as needlework is now.   The poet

Lace-making

Cowper thus described the lace-makers towards the end
of the eighteenth century :—

" Yon cottager who weaves at her own door,
    Pillows and bobbins all her little store;
    Content, though mean, and cheerful if not gay;
    Shuffling her threads about the livelong day:
    Just earns a scanty pittance, and at night
    Lies down secure, her heart and pocket light."

With the rise of machine-made Nottingham lace during
the nineteenth century the home-industry declined, but
of late years there has been a revival owing to the formation of associations to encourage the sale of hand-made

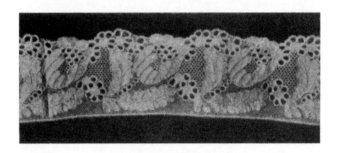

Buckinghamshire Lace

lace. The industry is carried on chiefly by old women
who learned it as children in the first half of the last
century, but a certain number of younger married
women carry it on in the time they can spare from
housework.

In a number of villages around Amersham and Wycombe, the women work at the ornamentation of bonnets
and dresses with beads and sequins.

The abundant production of milk in the pastures of the Vale of Aylesbury has led to the establishment of milk-condensing factories at Aylesbury. There are breweries in all the towns, but those of Marlow are perhaps the most noted. At Slough there are large nursery-gardens and seed-trial grounds, a factory for the making of a famous embrocation for horses and cattle, and engineering works.

The manufacture of needles was for several centuries an important industry at Long Crendon, but in the latter part of the last century it died out, the needle-makers gradually migrating to Redditch in Worcestershire.

13. Minerals.

The minerals that can be raised in any county depend on the rock-layers that form its subsoil. Thus chalk can be quarried in South, but not in North Bucks. The value of minerals depends upon a number of considerations—the depth at which they occur, and the ease with which they can be transported from the mine or quarry to where they are wanted, as well as their general usefulness. For instance there is a great area of chalk in South Bucks, but the statistics for 1907 record the raising of only 300 tons, and though this must be increased by the output from shallow pits which is not officially reported, the total is only a trifle as compared with the three million tons raised in Kent. This is because in

Kent much of the chalk lies close to the estuary of the Thames where it or its products—lime and Portland cement—can go straight into ships to be carried easily by sea; while Buckinghamshire chalk must travel by land at much greater cost. In South Bucks, therefore, chalk is only quarried for local use.

The flints, both in the chalk and in some of the superficial deposits, are used to some extent for building, as may be seen in the photograph of High Wycombe church on p. 128; and those picked off the fields are used for road-mending. More and more, however, the road-menders are coming to rely upon material from other counties such as Leicestershire and Warwickshire, where igneous and other hard rocks can be quarried on a large scale.

At Walters Ash, near Hughenden, the sandstone-blocks in the drift are systematically sought for as building-stone. By means of long iron rods thrust through the loose sand the position and size of the blocks are determined, and they are then dug out if large enough.

In the extreme north of the county the Oolitic lime-stones are quarried to a small extent for building-stone, but only for local use.

The most important mineral product of Buckingham-shire is clay, used for brick-making. There are two chief kinds of this clay—the brick-earth of South Bucks, and the stiff Oxford and Kimmeridge clay of North Bucks. The former is a natural mixture of clay with fine sand, similar to much of the silt laid down by a river where its flow is slackened; and the brick-earth of the Slough

district, which lies upon old Thames gravels, was certainly
deposited by the Thames itself. The brick-earth of the

Potter at work, Winchmore Hill, Amersham

Chiltern plateau, around Chesham, Wycombe, and the
Chalfonts, cannot have been formed in quite the same

way, but it was certainly laid down by water. These brick-earths are suitable, not only for making bricks, but also for tiles, drain-pipes, and flower-pots.

For several centuries the district around Chalfont has been noted for pottery, and many ancient diggings can be recognised here and there. Within living memory, the pottery from this district, carried in panniers slung across asses' backs, used to be hawked about the streets of Wycombe. During the last half-century the industry has greatly decayed, owing in part to the replacement of pottery by tin-ware and iron-ware for many domestic purposes, and in part to the cheapness with which pottery can be brought from North Staffordshire. The old potter shown still at work on p. 97 remembers seven kilns within a mile of the solitary remaining one at which he now works. It is the abundant production of bricks and tiles in former times that explains why old cottages in this district are all built of brick or brick and flint, and roofed with tiles instead of thatch.

The brick industry of the Slough district is of much more recent development, being a result of the opening of the Great Western Railway, by which or by the canal the bricks are conveyed to London and other places in the Thames valley. The growth of an immense population in a district where there is no building-stone has caused a great demand for bricks.

At Brill in North Bucks there have been potteries from the thirteenth century at least until within recent years; and just as the pottery industry was dying out, the brick industry developed through the opening of railway communication with London. The bricks are now made from clays rather lower in the geological scale and less sandy than those from which pottery was formerly made—Kimmeridge Clay at Brill, Oxford Clay at Woburn Sands and Calvert.

Among mineral products, mineral springs may be mentioned—springs, that is, which contain an appreciable amount of dissolved mineral matter, sufficient to give the water a taste, and sometimes a medicinal value. Of such there are few in Bucks, and none are of any repute now. The chief are those at Dorton, near Brill; Chalvey, nea Slough; and Dadbrook Hill, near Haddenham. The two first were visited as "Spas" about a century ago, when "taking the waters" was a favourite amusement among the leisured sufferers from real or imagined maladies.

14. History of the County.

Though we know from their remains that the ancient Palaeolithic people inhabited the valley of the Thames and that of the Misbourne, and that the Neolithic people, the bronze-using Goidelic Celts, and the later iron-using Britons all in turn inhabited the plain of North Bucks; and though the Romans made roads and built villas in the county, the first record in written history deals with the invasion of the Saxons, whose descendants still form the bulk of the population.

In A.D. 571, says the *English Chronicle*, "Cuthulf

fought with the Britons at Bedford and took four 'tuns' (fortresses), Limbury and Aylesbury and Bensington and Eynsham, and the same year he died." The order in which these places are named seems to indicate an invasion from beyond the Ouse towards the Thames, but historians seem generally agreed that it was in the opposite direction, Cuthulf being a West Saxon. However this may be, the strategic importance of the places named in relation to the Chiltern barrier and the great rivers is evident. Bedford as the lowest place where the Ouse could be forded, and Bensington which guards important fords over the Thames in Oxfordshire, were the keys of the river-entries to the great plain that stretches between them. Limbury is an earthwork fortress at the source of the Lea and guards the gap in the chalk ridge through which that river flows; while Aylesbury is on an eminence near enough to watch over the three gaps of Tring, Wendover, and Princes Risborough. Eynsham in Oxfordshire stands near the junction of the Isis and Evenlode, guarding both valleys.

For more than two centuries after this, history is silent about Buckinghamshire, and then, in A.D. 792, we read of Offa, the famous King of the Mercians, founding St Albans Abbey, and granting to it lands at Winslow and its neighbourhood. Nearly another century passes before, in 884, there is recorded a Witenagemot held by Aethelred, Duke of the Mercians, at Princes Risborough.

It is not until the Danish invasions that mentions

of our county in history come to be more frequent. In 921, the *Chronicle* tells how "between Lammas and Midsummer the Danish army broke the peace and went by night and came upon men unprepared, and took no

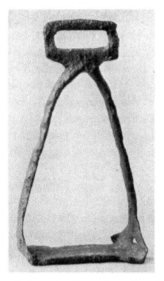

Danish Stirrup, ninth or tenth century, found at Longdown, Great Kimble

(*Aylesbury Museum*)

little, both in men and cattle, between Bernwood and Aylesbury." It was during this time of intermittent war between the Danes and the English that the Midland *shires*, including Buckinghamshire, were established. In the later Danish troubles, too, after Ethelred's massacre, Buckinghamshire was several times harried, as the *Chronicle* records for 1010 and 1015.

After the battle of Hastings, William the Conqueror marched through Buckinghamshire, along the foot of the Chilterns, to Berkhampstead, where he received the surrender of London.

On that Monday in June, seven hundred years ago, when King John coming from Windsor and the Barons marching from London met at the Thames, it was in Buckinghamshire "in the meadow which is called Runemed" near the Benedictine Priory of Ankerwyke that the Barons were encamped, and it was on an island within this county that John sealed the Great Charter which forms the foundation of English liberty (June 15th, 1215).

The Black Death visited Buckinghamshire in 1349, but we hear little of the peasants' revolt that in 1377 attained such importance in the neighbouring county of Hertford. In the religious upheavals that followed, however, Buckinghamshire took its fair share; in 1414, in 1452–72, in 1506, and in 1521 Wycombe, Amersham, and Chesham supplied Lollard martyrs to be burnt at the stake.

In the struggle between Charles I and the Parliament, Buckinghamshire played an important part. It was a Buckinghamshire squire, John Hampden, who, in 1637, refused to pay ship-money. When the Civil War broke out in the summer of 1642, one of the earliest battles was near Aylesbury, at the bridge where the Buckingham road crosses the Thame. While the King's headquarters

Magna Carta Island

were at Oxford, those of the Parliament were at Aylesbury, and much skirmishing took place in the ground between. Boarstall Castle and Hillesden House stood out for the King and were besieged, the former unsuccessfully, the latter successfully by the Parliamentary army.

Newport Pagnell also was occupied by the latter, to guard the lines of communication between London

Boarstall Castle

and the Midlands along Watling Street ; though it was temporarily taken from them by Prince Rupert in October 1643.

At Buckingham, in June 1644, Charles held the Council of War, from which he rode out to the battle of Cropredy Bridge.

Later historical events in the county centre round

the personalities of the Quakers of the seventeenth century, and the politicians of the eighteenth, and are recorded in Chapter 20. At the end of the eighteenth century, however, Buckinghamshire found itself strangely linked with French history by the fact that the exiled king, Louis XVIII, found at Hartwell House a temporary home for his court, as did the Orleans family later at Stowe.

Palaeolithic Implement found at Burnham
(*Aylesbury Museum*)

15. Antiquities.

Remains of the Palaeolithic race—their rough flint tools or weapons—are only found in Buckinghamshire in ancient gravels of the Thames and those of the Misbourne, as at Great Missenden. Just on the borders of Middlesex, in the Colne valley near Uxbridge, there was found in

1903 a vast number of flint implements and flakes (over 3000 from less than an acre of ground), which appear to belong to the very latest part of the Palaeolithic period.

Of the Neolithic race, who still used stone implements, but worked them smooth and polished them, there are found not only the tools, but also remains of their

Neolithic Celts

1. Taplow—not polished.
2. Babham Gulls, river Thames—moderately polished.
3. Haddenham—well polished.

(*Aylesbury Museum*)

habitations, and of the long mounds or barrows in which they buried their dead. Perhaps, too, some of the ancient roads and camps of England may be theirs. In Buckinghamshire their implements have been found chiefly in the Thames valley near Taplow, and on the Chiltern escarpment near Risborough. Remains of their hut-floors, dug out below ground-level, have been found at Hitcham and Ellesborough, but none of their long barrows are known in Bucks.

Following the Neolithic race came the Goidelic Celts,

Bronze Age Relics

1. Bronze Palstave, part of bronze-founder's hoard, found at Bradwell, 1879.

2. Bronze Armlet, found at Hartwell, 1848.

(*Aylesbury Museum*)

whose descendants still exist in the Highlands of Scotland, and who first introduced the use of bronze into England. They buried their dead in round barrows or tumuli, and some of the old roads, particularly the Icknield Way, may have been first laid out by this race. Hoards of bronze implements have been found at Bradwell, Waddesdon, Great Hampden, Hawridge, and along the Thames, and pottery of this age at High Wycombe, so that it is evident that the bronze-using people spread themselves over the whole county far more than did their predecessors. At Bradwell there was evidently a bronze-foundry, since broken objects had been kept with a view to re-melting.

Next came the Brythonic Celts, who introduced the use of iron into this island, and who were still the dominant race when the written history of Britain began

Iron Currency-Bar (Early Iron Age) found in the Thames
(*Aylesbury Museum*)

with the invasion of Julius Caesar. Iron tools and weapons being far more destructible than those of bronze, these remains are rarer; but an iron currency-bar of this time, dredged from the Thames, is shown above. The coins of the later Celtic period were of gold and copper, and they have been found all over the county, as at Chalfont, Chesham, Creslow, Cuddington, Drayton Beauchamp, Ellesborough, Fleet Marston, Quainton, Stoke Mandeville, Fenny Stratford, Thornborough, Wendover, Whaddon, and High Wycombe. The designers of this coinage copied a Greek coin of Philip

of Macedon on which was a beautiful representation of a
chariot and horses; but this design was copied and re-
copied until its original meaning was forgotten, and the
horses' legs and chariot wheels came to be strangely
mixed up, as may be seen below. Many of the
coins bear the names of British kings, such as Cunobelinus (Cymbeline), Tasciovanus, and Andocomius.

Ancient British Gold Coins found at Whaddon Chase, 1849
(*Aylesbury Museum*)

Pottery of this age has been found at Aston Clinton
and Bierton, silver ornaments at Castle Thorpe, and
an iron sword in a bronze scabbard near Taplow.

Although written history begins for us with the
Roman invasion, yet it has for many centuries to be
supplemented by the "unwritten" history inscribed on

the surface of the land. Many monuments exist of which we do not know the age.

Such are the two giant crosses cut on the Chiltern

Whiteleaf Cross, Monks Risborough

Hills by the simple process of removing the turf and exposing the bare chalk. The Whiteleaf Cross at Monks Risborough, which can be seen for many miles, has a large triangle at its base, and is apart from this 80 ft.

in height and 80 ft. in breadth. Bledlow Cross is smaller (70 ft. in height and breadth) and lacks the triangle, so that it is a much less conspicuous object. As crosses, these can hardly date back earlier than the conversion of the West Saxons to Christianity in the seventh century ; but it is believed by many that they

Lynchets, West End Hill, Cheddington

represent an alteration of earlier pagan symbols, cut long before, perhaps by the Britons, who are almost certainly the makers of the white horses cut elsewhere in similar situations. More sceptical critics, noticing that the antiquaries of the sixteenth and early seventeenth centuries make no mention of these crosses, and that there is no reference to them in any ancient documents, have doubted their antiquity altogether, and suggested that they may have been cut by the soldiers in the Civil War.

Scarcely less conspicuous, but the unintentional result of long years of agriculture, are the lynchets, linces, or cultivation-terraces on the outlying chalk hills of Cheddington. They were produced at a time when hill-side fields were ploughed horizontally, not up-and-down as is done now. Unploughed strips were also left at intervals, and as the sod was always turned downhill, not uphill, in course of many years each ploughed strip of hill-side became a nearly level terrace, with an accumulation of soil towards its lower unploughed boundary. Similar lynchets, there called balks, are to be seen on the hill-side east of Chesham. Which of the successive races of men produced these terraces we do not know.

A remarkable earthwork of unknown date is the Grims-dyke or Grim's Ditch, which runs along the Chiltern plateau, roughly parallel to, and about a mile away from, the escarpment. It is a mound or wall of earth having along its south-eastern side the ditch from which was dug out the material to make it. It may have been intended as a fortification, but it is difficult to understand how a rampart of this length could have been guarded along its course of many miles; and it is possibly simply a tribal boundary. If so, it must belong to a period earlier than that of the Saxons, as their parish-boundaries pay no attention to it. The name, Grim's Ditch, must have been given by Saxons who thought it of superhuman origin, the work of Giant Grim.

Grimsdyke, near Lacy Green

The principal Roman remains in the county are at High Wycombe, Terricks (near Ellesborough), Latimer, Shenley, Stony Stratford, Foscott, and Tingewick, at all

Roman Silver Vase and Spoons

Found between Great Horwood and Winslow, in 1872. One of the spoons has engraved in the bowl "Veneria vivas."

(*Aylesbury Museum*)

of which places the remains of buildings with tesselated pavements have been found. The station of Magio-vintum on Watling Street seems to have been at Little

Brickhill, and many miscellaneous Roman remains—coins, ornaments, etc.—have been found there. Burial-places have been found at Long Crendon, Stone, and Weston Turville. Miscellaneous finds of coins and pottery have occurred at many places scattered all over the county.

In many parts of the county we find "Camps," that is to say areas varying from a dozen acres downwards, surrounded by an earthen wall and ditch to protect them from attacks by an enemy. Of course, the ditch is outside the rampart. Some of these camps are square in shape, others more or less circular, and it was at one time thought that all the former were of Roman origin, and the latter British, Saxon, or Danish. This is not however by any means certain. The best example of a square camp is that on Muswell Hill, on the Oxfordshire boundary. There are others at Shenley, Burnham Beeches, and a very fine one (700 ft. by 500 ft.) at Bow Brickhill, known as Danesborough.

The finest round camp is that at Cholesbury, nearly 1000 feet in its longer diameter, within which the church of St Lawrence is built. There is another, with an area of 21 acres, in Bulstrode Park.

Tumuli, or mounds raised over the burial-place of some important person, are not uncommon in certain parts of the county, and were probably commoner once because many have been levelled for agricultural purposes. There are no long barrows in Bucks such as the Neolithic people used to raise, and the round barrows that are found may belong to any period from the Bronze Age to the pagan period of the Anglo-Saxons. Many have

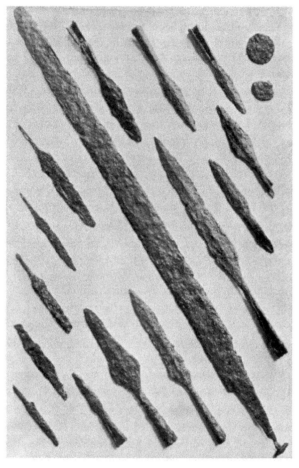

Old English Weapons, found at Bishopstone, 1866

(*Aylesbury Museum*)

yielded remains clearly belonging to the last of these, as in the case of the great tumulus of Taplow, on which the old church was built and an old yew tree had grown to a girth of 20 feet. The greatest number of Saxon

Old English Drinking Glass found in Cemetery,
Newport Pagnell, 1899

(*Aylesbury Museum*)

burials have been found in the district between Aylesbury and Thame—at Bishopstone, Hartwell, Stone, Dinton, Winchendon, Ashendon, and Kingsey. Not in all cases were tumuli visible above ground when the burials were unearthed. There are other Saxon cemeteries at Wycombe and Newport Pagnell. In pagan times persons were buried fully dressed with weapons, ornaments, drinking-cups, etc. alongside them, so that from these we can learn a great deal of the arts and customs of those

Thornborough Mounds
(*Romano-British tumuli*)

days. Examples of such objects from these burial-places are shown on pp. 116 and 117.

The Thornborough mounds, shown above, belong to an earlier period, the remains dug out from them being of Romano-British character; and there are tumuli at Cublington, Salt Hill near Slough, and elsewhere of

Denham Lodge, Quainton

(The only inhabited moated house in Bucks)

whose age we know nothing. The greatest abundance of tumuli, however, is on the bare Chiltern downs, at Bledlow and Saunderton and Ivinghoe. Perhaps one of those at Bledlow may conceal the remains of the forgotten Saxon named Bledda who gave the place its name— Bleddan hlaew (Bledda's tumulus).

Moats, that is to say, ditches filled with water originally surrounding a house, are so common in the clay-plain north of the Chilterns that it would be wearisome to give a list of them. There are about thirty at least, but in nearly all cases the house which they once surrounded has gone. Denham Lodge, near Quainton, is the only example in Buckinghamshire of a house still inhabited, surrounded by a moat. A very few similar moats are also found on the Chiltern plateau near Chesham.

16.   Architecture—(a) Ecclesiastical.

The study of early architecture in England is mainly a study of churches, because these were built of stone at a time when very few other buildings were, and because they have been less frequently destroyed than other stone buildings, however much they may have been altered. It is necessary therefore to have a clear idea of the construction of a church in order to understand the early history of English architecture.

The simplest churches are oblong buildings with their length lying east and west. The entrances are either at the west end or in the north and south sides. Inside, the church is divided into a larger portion, the *nave*, and a smaller *chancel* at the eastern end, separated generally by a *chancel arch*. At the east end of the chancel is a window, in front of which stands the altar. Simple churches of this kind are seen at Foscott and Fleet Marston. At Grove there is not even a chancel arch. There is often a *tower* at the west end. In early times this was probably built as an outlook and a place of refuge in time of war: afterwards they chiefly served to hang the bells, and when architecture became a fine art the possibilities of beauty in a tower and spire led builders to devote great care to their construction. Simple churches with towers exist at many places—among others, Adstock, Akeley, Dorney, Water Stratford, Grendon Underwood, Broughton, Linslade, and Hitcham.

Usually a church has in addition to tower, nave, and chancel, an *aisle* along one or both sides of the nave. When a simple church had aisles added to it, this was done by knocking holes in the nave-walls, and fitting arches to them, or by taking them down entirely and replacing the wall by an open arcade, and building new aisle-walls on the north and south sides. To make good the loss of light to the nave, a clerestory (explained below) might be added. When a church was built originally with aisles, these are only separated from the nave by a series of pillars and arches. Sometimes the nave and aisles are covered by a roof of uniform slope, as at Bierton and Wraysbury. At other times the nave is much higher than the aisles, and the side-walls above the aisle-arches form a *clerestory* with windows, as at Waddesdon, Padbury, and most of the larger churches.

Though much detail might be added, the above will serve to explain the principal features in any church.

The oldest style of architecture in this country is commonly known as Saxon or pre-Norman, and belongs to the period before the Norman conquest. This was the work of early craftsmen with an imperfect knowledge of stone construction, who commonly used rough rubble walls, no buttresses, small semicircular or triangular arches, and square towers with what is termed "long-and-short work" at the quoins or corners. It survives almost solely in portions of small churches.

The most important pre-Norman structure in Buckinghamshire is part of the church at Wing, including a crypt or underground chamber beneath the chancel—one out of the only six known to exist in England. These crypts were safe hiding-places for the sacred relics, which played so important a part in religious ceremonies in early days. Like other crypts that at Wing consisted of a central relic-chamber, with a passage around it along which the people walked in procession and which communicated with the church by a flight of steps. Now, however, this communication with the church is stopped up, and the crypt can only be viewed from outside. One peculiar feature resulting from the presence of a crypt is that the east end of the church is rounded instead of square as usual; such a rounded end is called an *apse*.

No other undoubted pre-Norman structure exists in the county, though a good Saxon tower remains at

Caversfield, which was formerly a detached part of Bucks, but has long been included with Oxfordshire.

The tower at Lavendon is certainly very old, and its very rough masonry, and almost childish attempts at

Saxon Apse, Wing

(Two of the crypt-windows are seen at ground-level. The round and triangular arches are Saxon. Windows have been inserted at a much later date)

arches, suggest a pre-Norman date, but it does not show any of the distinctive features of Saxon work, and we must remember that rough work might have been done in one place at the very time when the beautiful Norman work was being done elsewhere.

There are some cases, however, in which late Norman work can be seen to have been done in alteration of older work, and it is not unlikely that that older work may be earlier than the Norman Conquest.   Thus at Iver in the

Stewkley Church, from the north-west

(An almost unaltered late Norman church.   Even the high-pitched roof has been left)

south of the county, and Lathbury in the north, the church seems to have at first consisted of a simple nave or body, with a chancel at the east end, but no aisles. Afterwards, archways were cut in the side-walls of the nave, pillars inserted and the tops of old windows blocked

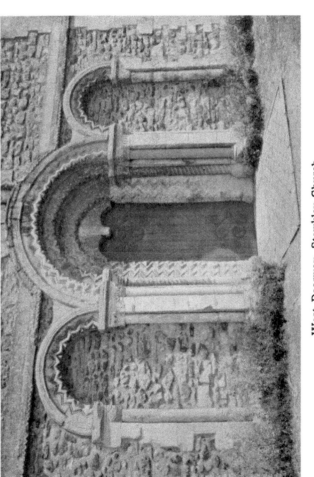

West Doorway, Stewkley Church

(*Showing the characteristically Norman zigzag ornament on the arches, and typical shape and ornament of the capitals and pillars*) up, while aisles were built on. The remains of the old walls and their blocked-up windows are still to be seen, and *may* be of Saxon date.

The Norman conquest started a widespread building of massive churches and castles in the continental style called Romanesque, which in England has received the name of "Norman." They had walls of great thickness, semicircular vaults, round-headed doors and windows, and often lofty square towers. The capitals which topped the pillars and supported the arches were square, and they and the arches were ornamented in various designs not found in later styles.

The Norman style is illustrated in its very plainest and simplest form by the church at Little Missenden, which has escaped any serious alteration from the twelfth century; and in its highest development by the beautiful church of Stewkley, which has also been very little altered. In these two cases we can see a church very much as it was built in the Norman period; but there are many other cases where single Norman doorways have survived in buildings otherwise rebuilt, as in the churches at Dinton, Horton, Upton, Water Stratford, and many others; and in St John's Chantry at Buckingham. This is because the Norman doorways were good enough to satisfy later architects, who found Norman windows much too small.

From 1150 to 1200 the building became lighter, the arches pointed, and there was perfected the science of vaulting, by which the weight of the roof is brought upon piers and buttresses. This method of building, the

South Window, Chetwode Church

(Early English, with thirteenth century stained glass)

South Porch, High Wycombe Church

(Early English—the wall can be seen to be largely built of flint, and the flint and stone chequer-work over the window is very evident)

"Gothic," originated from the endeavour to cover the widest and loftiest areas with the greatest economy of stone. The first English Gothic, called "Early English," from about 1180 to 1250, is characterised by slender piers (commonly of marble), lofty pointed vaults, and long, narrow, lancet-headed windows. After 1250 the windows became broader, divided up, and ornamented by patterns of tracery, while in the vault the ribs were multiplied. The greatest elegance of English Gothic was reached from 1260 to 1290, at which date English sculpture was at its highest, and art in painting, coloured glass making, and general craftsmanship at its zenith.

After 1300 the structure of stone buildings began to be overlaid with ornament, the window tracery and vault ribs were of intricate patterns, the pinnacles and spires loaded with crocket and ornament. This later style is known as "Decorated," and came to an end with the Black Death, which stopped all building for a time.

An interesting example of the transition from Romanesque to Gothic is seen in the chancel arcading at Wingrave. Here the pillars have square capitals with ornament of Roman character, but support plain pointed arches.

Of typical thirteenth-century or Early English style, Chetwode furnishes a very beautiful example; and some of the beautiful contemporary coloured glass is preserved here. There are single doorways of this style at Cuddington, High Wycombe and elsewhere.

Milton Keynes church is a very perfect example of a "Decorated" church, unspoiled by later styles.

D. B.   9

East Window, Emberton Church

(*This illustrates the highest development of tracery, characteristic of the Decorated style*)

other beautiful examples of this style are the churches of
Preston Bissett and Emberton.

With curious uniformity and quickness the style
called "Perpendicular"—which is unknown abroad—
developed after 1360 in all parts of England and lasted
with scarcely any change up to 1520. As its name
implies, it is characterised by the perpendicular arrangement of the tracery and panels on walls and in windows,
and it is also distinguished by the flattened arches and the
square arrangement of the mouldings over them, by the
elaborate vault-traceries (especially fan-vaulting), and by
the use of flat roofs and towers without spires.

Two very fine examples of this style are to be seen in
North Bucks. Maids' Moreton church, near Buckingham,
was entirely rebuilt in 1450. The east window (p. 132)
shows a feature that at once marks the style, because it
would have been quite impossible to a "Decorated"
architect—the carrying up vertical lines in the tracery
both to intersect the curve of an arch and to support
the point of an arch. This shows that these arches
are purely ornamental, not essential to the construction.
There is fan-tracery in the porch over the western
doorway.

Hillesden church was built even later (about 1493).
Sir Gilbert Scott, a great admirer of the style, described
it as "the choicest specimen of a village church in the
county: very few in England, of its period and scale,
surpass it."

In South Bucks the chapel of Eton College is a very
fine example of the Perpendicular style.

East Window, Maids' Moreton Church

(*Showing characteristic Perpendicular window*)

Of more modern churches, the only one worth special mention is that of Willen, built by Sir Christopher Wren, the architect of St Paul's in London.   Unlike the churches of the middle ages, it does not look beautiful to our eyes,

Hillesden Church

(The flattening of the arches and predominance of vertical bars in the window tracery show this to be late Perpendicular in style.   Remains of the churchyard cross in the centre)

whatever may have been thought about it at the time when it was built (about 1680).   The failure of so great an architect to produce a beautiful church may be ascribed

Eton College Chapel

(A very fine example of the Perpendicular style, dated about 1442)

in part to the fact that he was too busy rebuilding London churches after the great fire, and hardly considered the difference between a village church and one standing between the houses of a city street ; but it is in large measure due to the unwise attempt to build in imitation of the ancient buildings of Rome and Greece a church, that is to say a building whose design had been evolved under quite different conditions.

Buckinghamshire has few other ecclesiastical buildings than its parish churches. It has never had a cathedral, nor did it contain any great or ancient monasteries or abbeys during the middle ages. There were a few monasteries, but of most of these nothing remains.

It was not until the end of Henry I's reign that the first abbeys and monasteries were founded in the county. Then Missenden Abbey and Tickford (Newport Pagnell), Luffield, and Ivinghoe Priories were founded. Of these only part of Missenden Abbey remains, incorporated in a dwelling-house. In the latter half of the twelfth century there were founded Bradwell Abbey, Ankerwick Priory (by the Thames), Notley Abbey (by the Thame), and Lavendon Abbey. Of these, only some fragments of Notley remain, Early English work incorporated in modern farm-buildings. Medmenham Abbey, on the Thames, was founded at the beginning of the thirteenth century, and of this there are the best of any remains, of beautiful Early English style. Snelshall, Chetwode, and Ravenstone Priories, Burnham Abbey, Bulstrode Preceptory, and Ashridge College were founded in the thirteenth century, and of these all that remains are part

Medmenham Abbey

(The lower arches are plain Early English ; the tower windows show one of the earliest forms of tracery. The doorway and windows are of much later date)

of the domestic buildings of Burnham Abbey, and of Snelshall Priory. The only Franciscan Friary in the county, at Aylesbury, was established in 1386.

17.   Architecture — (*b*)   Military   and Domestic.

Buckinghamshire is strikingly deficient in the remains of military architecture. Only one portion of a castle actually remains standing, and that is of comparatively late date—the gatehouse of Boarstall Castle (p. 104), built in the fifteenth or sixteenth century, and besieged in the Civil War.

The earliest castle built in the county was probably that of Buckingham, on a hill surrounded by a horse-shoe loop of the Ouse; but of this not a vestige remains, and its site is only attested by the name of Castle Hill, where the modern church stands.

Of a number of other castles, some of them built probably in the reign of Stephen, often on the site of older earthworks, only the earthworks now remain. Such are Lavendon Castle in the extreme north of the county, Hanslope Castle or Castle Thorpe, Bolbec Castle on a height overlooking the Vale of Aylesbury at Whitchurch, Desborough Castle near Wycombe, and Hedsor Castle near the Thames. The so-called Dinton Castle is a little tower built for no serious purpose in the eighteenth century.

Few old houses remain in an unaltered state in the

county. Perhaps the best example is the old Court-House or Staple Hall at Long Crendon, if it be a fourteenth-century building as is believed. It is a half-timbered construction, originally built perhaps for a wool-store and afterwards used, in the fifteenth century, for the holding of manorial courts. Though now modernised by the

Court-House, Long Crendon

insertion of windows to render it habitable, it shows the fine old beams, and the upper storey is little altered. It now belongs to the National Trust for Places of Historical Interest, and though let as a residence may be visited by the public.

Portions of fourteenth-century manor-houses remain

Windows at the King's Head, Aylesbury

at Creslow (where parts of the adjacent Norman church are used as farm-buildings), Doddershall House (near Quainton) and Hampden House, but these have all been greatly altered in later times. The King's Head Inn at Aylesbury (p. 139) was built about the middle of the fifteenth century, and its windows still retain some of the contemporary glass. A fifteenth-century barn is preserved at Great Kimble.

The dissolution of the monasteries and the distribution of their wealth among private landowners led to an outburst of manor-house building about the years 1530–1540. The style of this period is known as "Tudor," and marks the final stage of Gothic architecture when pointed arches became lower and lower until they passed into the modern flat-headed window. With the change in the main object of architecture from the church to the dwelling-house, ceilings and panelling came in, and chimneys came to be a feature to be treated ornamentally. The finest example of this style in Bucks is Chenies manor-house, built by Lord Russell about 1540. Originally built about a quadrangle, only the south side now remains, with the curious result that all the windows are on the sunless north side. The main buildings of Eton College are of this style, though of somewhat early date.

Smaller examples are the manor-houses at Marsh Gibbon and Buckingham; the latter with a twisted chimney. Chequers Court and Gayhurst House are, in part, of the same or slightly later date, but both were greatly altered in the eighteenth century. The Old Rectory House at Beaconsfield, now in use for educational

Eton College

(Tudor style, about 1517)

purposes, is a brick and timber building of early sixteenth-century date.

A later development of the same style was the Elizabethan, represented by Latimer House, Fullbrook House, and Hartwell, though the last-named was greatly altered in 1749.

Gayhurst
(*An Elizabethan mansion*)

With the seventeenth century foreign influence brought in the characteristic ornaments of the "Renaissance," derived from the revival of the study of Greek and Roman antiquities. The round arch was re-introduced after being discarded for four centuries, but the Renaissance arch differs from a Norman arch, not only in ornament but also by the projection of the key-stone.

The grafting of the Renaissance style on the Elizabethan produced in England what is called the Jacobean, of which we have examples in Castle House, Buckingham, and the Drake Almshouses and Raans manor-house near Amersham.

Of later date in the seventeenth century (after the

Claydon House and Middle Claydon Church

Commonwealth) but of generally similar style are Quainton Almshouses, the Town Hall of Amersham, and many still-surviving dwelling-houses such as Little Shardeloes at Amersham.

Cliveden House, near Taplow, was built by George

Villiers, Duke of Buckingham, but shows the character-
istics of the "classic" style of the eighteenth century, in
which the last trace of Gothic influence—the mullioned
window—had disappeared. The finest example of this
style is Stowe House, the front of which was completed
by Earl Temple in 1760–80. Other examples are
Shardeloes House, Amersham, and Claydon House,
though the last contains a Tudor core.

Of later styles the examples are too numerous to
mention. Many of the cottages of the county are of
eighteenth-century date. Local conditions naturally
determined the details of construction of these cottages,
especially as regards the materials of which they are constructed. Thus in South Bucks, where flint and brick-
earth abound, the old cottages are of brick, or of flint
walls with brick corners, and are almost invariably tiled.
In North Bucks they are built of brick and timber, and
are almost invariably thatched.

18. Communications—Past and Present. Roads, Canals, and Railways.

From the time when men first began to use metals
instead of stone for their tools and weapons, if not before,
some means of conveying goods from place to place have
been necessary; and in England, even before the Romans
introduced their civilisation, roads were already in existence. One of the chief of these connected the mining
districts of the West of England with the rich agricultural

districts of the Eastern Counties. The traders of those days were guided in making their tracks by the conspicuous physical features of the country, for they had neither maps nor surveying instruments. They avoided as far as possible the wooded and swampy clay-plain, and kept to the edge of the chalk escarpment. In Berkshire, where

Icknield Way at Whiteleaf

the crest of the escarpment is less broken by valleys and there are bare downs instead of beech-woods, this road keeps high up on the crest; but in Buckinghamshire it runs near the foot of the steep slope of the escarpment.

It is known as the Icknield Way, and in parts of its course it divides into two parallel roads—the Upper and Lower Icknield way. This road has long ceased to be of

importance for long-distance traffic—ever since London
became important enough to divert trade into its own
channels; and in places the track has been lost; but much
of it still survives, sometimes as a local, second-rate road,
sometimes as a green track-way or bridle-path.   It enters
the county near Bledlow, passes through the Risboroughs,
the Kimbles, Wendover, and Ivinghoe, and leaves it near
Edlesborough.

It says much for the former use of this road and
similar roads following the course of the chalk escarpment,
that two plants that grow only on chalk and limestone
should have been named the *traveller's joy* and the *way-
faring tree*.   Those names could hardly have come into
use among travellers who followed roads that struck across
chalk-ridge and clay-plain alike ; but to many a dweller
on a clayey or sandy soil the abundance of these plants
must have been a striking feature of a journey along the
Icknield Way.

Among other ancient roads of unknown date is the
Angle Way at Long Crendon, which has a Roman
cemetery near it, and is a continuation of the road leading
westwards from Aylesbury, along which are several heathen
Saxon burial-places.   There is also Weasel Lane, which
runs from Fenny Stratford to Winslow, and forms a
parish-boundary for much of the way—a sure sign of
antiquity, at least in a road that is fairly straight, for it
shows that the road probably existed before the parish.
If, however, the road twists and turns, it is more probable
that it grew up as a right-of-way along the boundary of
a township already marked out.   Many parish-boundary

lanes exist in the county, and these are often green lanes, terribly muddy in winter.

Some of these were habitually taken advantage of by the Welsh drovers who, to within living memory, drove their herds of sheep and cattle up to London or to various fairs, and they are spoken of as Welsh Lanes.

The Romans made the finest roads that have ever been made in England. Their great road from Dover through London to Chester, which in Saxon times was given the name of Watling Street, crosses the Chilterns at Dunstable, some way east of the Bucks border, and only enters our county among the sand-hills of the Bedfordshire border. Brickhill, the first Buckinghamshire village on this road, acquired some importance from this fact in medieval times, as the justices in circuit stopped there to try Buckinghamshire criminals. The road keeps within the county for only 10 or 11 miles, crossing the Ouzel at Fenny Stratford and the Ouse at Stony Stratford. Throughout this distance it is still a great main road, well known to cyclists and motorists.

Another Roman road, called by the Saxons Akeman Street, running originally from Verulamium (St Albans) to Aquae Sulis (Bath), passes through our county. It crosses the Chilterns by the Berkhampstead gap, covered by the intrusive wedge of Hertfordshire, and enters the county near Drayton Beauchamp. Passing through Aston Clinton, Aylesbury, and Waddesdon it leaves the county near Grendon Underwood, in which village Shakespeare is said to have slept one midsummer night, when on a journey between London and Stratford-on-

Avon, and according to local tradition conceived the *Midsummer Night's Dream*. Akeman Street is still in use as a great main road, but its St Albans end and much of its Oxfordshire portion have been lost, and it now takes motors from London to Birmingham and other places in the Midlands.

The only other certain example of a Roman road in Bucks has had a worse fate. Leading originally from Dorchester on the Thames to Towcester, or near it, only fragments of it survive. One of these forms the county-boundary for a mile near Barton Hartshorn and Finmere. That it once continued in a straight line thence to the ford at Water Stratford is evident from the fact that the county-boundary does so continue, though the road is gone. On the other side of the Ouse it can be traced as a minor road across Stowe Park, after which it is again lost.

Many of the minor roads of the county may possibly be as old as these or older, but it is impossible to prove it. After the English conquest, no roads were made for long distances, but tracks from village to village came to be utilised as long-distance routes as trade increased. That is why so many of our main roads twist and turn in passing through a village. Nor are they by any means straight between the villages. The road from Aylesbury to Leighton Buzzard, for instance, makes a series of right-angled bends where there are few or no houses. We can only guess that this is because it came into being at a time when agriculture was more important than trade, the rectangular shape of cornfields a more urgent matter than the straightness of a roadway. During the eighteenth century the demands of growing trade and the introduction of stage-coaches compelled improvements to be made in roads, and many of the worst twists must have disappeared then; but the advent of railways put a stop for a long time to the improvement of the roads.

In the time of stage-coaches, the Bath Road was as important as it has now again become since the advent of motors. It is probably in parts an ancient road, but with many portions made in comparatively modern times. It enters the county at Colnbrook, passes through Slough, and leaves it at Maidenhead Bridge.

For many years Buckinghamshire was badly served with railways. When George Stephenson was planning the London and Birmingham railway—the first long-distance line, now merged in the London and North-Western— he proposed to cross the Chilterns by the Wendover gap, and to make Aylesbury and Buckingham stations on the main line, but the Duke of Buckingham disliked the idea of a railway close to Stowe House, and with other landowners drove Stephenson to take the Berkhampstead gap, thus preventing Buckingham from becoming what Bletchley is now. Consequently the main line only runs through the north-eastern part of the county, instead of through the centre. It happened however that Wolverton was about half-way between London and Birmingham, and two things, as already stated on a previous page, resulted from this. First, for many years all trains stopped there for refreshments; and secondly, works for building and repairing the locomotives were set up

The London and North-Western Railway Co.'s Carriage Works, Wolverton

there. When the London and Birmingham Company amalgamated with others to form the London and North-Western, all locomotive building was transferred to Crewe, and the Wolverton works were used exclusively for building railway carriages—an industry of very great importance there now.

The Aylesbury people did not share the landowners' objection to railways, and constructed a branch to their town at quite an early date. Afterwards, lines to Oxford and to Cambridge were constructed from Bletchley, which then took the place of Wolverton as a stopping-place for expresses.

The Great Western Railway from London to Bristol, constructed by Isambard Brunel, follows the Thames valley, entering Buckinghamshire near West Drayton, and leaving it again at Maidenhead Bridge. Slough, the junction for Windsor, is the most important station on these 10 miles of line, which are in the extreme south of the county. A branch from Maidenhead through Wycombe to Aylesbury and Thame was constructed in the days when capital was difficult to raise, and it was therefore roundabout, with sharp curves and severe gradients, and a single line with very few crossing places.

The branch of the London and South-Western Railway from Staines to Windsor lies mainly within the county, in the extreme south, and the Midland branch from Bedford to Northampton serves Olney in the extreme north. For many years these were the only railways in Buckinghamshire except two small, local lines—one from Aylesbury to Verney Junction and the other from this line at Quainton Road to Brill. Both these lines were worked in a very primitive way. As recently as 1897 the guard on the Brill train might be seen standing up on an open truck and throwing out empty milk-cans on to the side of the line as the train passed a farm-house.

Within recent years two new main lines have been constructed through the heart of the county. The first passes through Amersham (with a branch to Chesham), and by the Wendover gap to Aylesbury, then to Quainton Road over the route of the local line above mentioned, and then across the plain to Brackley (in Northampton-shire, just outside our county) on its way to Leicester and Sheffield. This is the Great Central Railway, the portion as far as Quainton Road being jointly owned by the Metropolitan Railway Company.

The second belongs jointly to the Great Central and Great Western Companies. It enters the county at Denham, near Uxbridge, and makes a nearly straight line to Wycombe; then along the old branch line (improved in curves and gradients) to Princes Risborough, and from there through Haddenham to near Ashendon, where it divides—the Great Central branch joining the line already described near Grendon Underwood, while the Great Western line runs through Brill and Ludgershall on its way to Bicester and Birmingham.

Before the time of railways, in the eighteenth century, many canals were made, but only one of these passes through Buckinghamshire—the Grand Junction Canal.

This follows approximately the route of the London and North-Western Railway, but with differences. A canal is much more dependent on the physical features of the land than railways, since it can only ascend or descend a slope by means of locks, which are a hindrance to traffic. A canal will therefore make considerable detours to keep nearly level, and the Grand Junction Canal starts from London by the route of the Great Western Railway, which a branch of the canal follows to Slough, the main canal turning northwards along the Colne valley—a small portion of this part of its course being in Bucks and the rest of it just outside the county. After a course through Hertfordshire it re-enters our county along with the London and North-Western Railway near Marsworth. This is just at the highest point, or summit level, of this part of its course, and to avoid the danger of the highest reach being left dry, since the water runs out from it both ways, there are here large reservoirs, though, as it happens, they lie almost entirely in Hertfordshire. A branch extends from here to Wendover, and another to Aylesbury, to which place it descends by 17 locks.

From here to Wolverton it follows the railway, but with so meandering a course that what the latter performs in 20 miles the canal takes over 30. A branch to Buckingham runs along the north side of the Ouse and is therefore for part of its course in Northamptonshire.

19. Divisions—Ancient and Modern. Administration.

It was explained at the beginning of this book that counties which take their name from a town were prob-ably made a thousand years ago by combining a number of the smaller divisions known as hundreds. In the case of Buckinghamshire we know from Domesday Book that there were eighteen hundreds in six groups of three each; and later these groups came to be reckoned as single hundreds, called, for the most part, after the principal town in each. In the following list the eighteen names are those found in Domesday Book, while the modern name, if there is one, is placed in brackets after it.

1.	Stanes (Stone)	} The Three Hundreds of
2.	Elesberie (Aylesbury)	Aylesbury.
3.	Riseberg (Risborough)	(Counted as one.)
4.	Stoches (Stoke)	} The Chiltern Hundreds.
5.	Burneham (Burnham)	(Stoke, Burnham and Des-
6.	Dustenberg (Desborough)	borough, counted as three.)
7.	Ticheshele (Ixhill)	
8.	Essedene (Ashendon)	} Ashendon Hundred.
9.	Votesdone (Waddesdon)	
10.	Coteslai (Cottesloe)	
11.	Erlai	} Cottesloe Hundred.
12.	Mursalai (Mursley)	
13.	Stodfald	
14.	Rovelai	} Buckingham Hundred.
15.	Lamua	
16.	Sigelai	
17.	Bonestou	} Newport Hundred.
18.	Molesoveslau (Moulsoe)	

It is chiefly in the northern part of the county that the old names of hundreds have been lost.   In the southern part, they are nearly all the names of villages—doubtless the villages in which the hundred-moot or court was held.   But in the northern area many of the names seem to take us back to a more ancient state of things—the meeting of the freemen of the hundred, not in a village but in some wild open place.   Thus six names end in *lai* or *lau*, which in most cases, though perhaps not at all, means *hlau*, a tumulus; and we may imagine the meeting taking place by some hero's funeral mound.   Stodfald means a *stud-fold*, that is, an enclosure for horses; and Ixhill is not near any village.

In later times these hundreds were found to be too small and numerous for convenient administration, and they were grouped in threes and the court held at the chief town in each.   The only exception was in the extreme south of the county, where, owing to the few settlements made in early times in the Chiltern Forests, the hundreds were much larger than in the rest of the county.   They therefore remained distinct, and so in modern times there were reckoned to be eight hundreds in all in the county (the Three Hundreds of Aylesbury counting as one).

The business of the hundred courts was to levy taxes, to look after roads and bridges, and to try criminals and fine them, for even a murder could in early days be expiated by a fine.   The steward of a hundred collected the taxes and fines, and either paid them over to the king or the earl of the shire, keeping a commission for himself, or took the stewardship "on farm," paying an agreed sum as rent and keeping all that he could collect. The importance of the hundred court was sapped in several ways as time went on. The lords of manors, the justices of the peace, and the king's justices on circuit took more and more of the criminal jurisdiction into their hands, and the sheriff took charge of the taxation. So that in most cases the hundred courts long ago became obsolete by having nothing to do, and they were finally abolished in 1867. The areas of the hundreds, however, are still in use by the magistrates as Petty Sessional divisions.

The three Chiltern Hundreds have a peculiar interest from their curious modern association with Parliament. The House of Commons, in 1623, passed a resolution "that a man, after he is duly chosen, cannot relinquish" his membership; and this rule, passed at a time when it was often necessary to compel a county gentleman to represent his shire in Parliament when he would have preferred to stay at home, still continues in force. But by a statute of 1707, any member of the House of Commons who accepts "any office of profit from the Crown," at once ceases to be a member, but is capable of being again elected. This is the familiar case with members who become cabinet ministers, and are obliged to seek re-election.

It happened, however, in 1750, that Mr John Pitt, M.P. for Wareham, wished to vacate that seat and become a candidate for Dorchester. Not being able to resign, he sought for some office of profit under the Crown which had no real duties or pay attached to it, and yet would

necessitate vacating his seat.   He hit upon the Stewardship of the Chiltern Hundreds, and from that time
onwards this nominal office has been in use for no other
purpose than that of enabling members of Parliament to
evade the resolution of 1623.

The duties of the Steward of the Chiltern Hundreds
were to hold the Hundred Courts, and receive the various
rents, fines, fees, tolls, etc., due to the Crown from the
men of the hundred.   In the thirteenth and fourteenth
centuries these amounted to a considerable revenue, and
the Steward's office was a very profitable one; but, as
already explained, the importance of the hundred courts
was ever diminishing.   In the case of the Chiltern
Hundreds, the Stewardship was farmed out in 1679 for
31 years at a yearly rent of £1. 6s. 8d. and a fine of £150
—the annual income being £18. 6s. 8d.   That is the
last that is recorded of the profits of this Stewardship,
although a few of the rents, amounting to less than five
pounds in all, are still collected by the Treasury to this day.

Just as the county was made up of a collection of
hundreds, so each hundred was made up of a collection
of townships.   A township was a settlement of a number
of families living close together, and its area was the
total of the fields, meadows, common pasture, and woodland which they made use of in their agriculture.   Such
an area was not always one compact piece—there might
be bits of woodland, perhaps, at a distance from the other
lands, but belonging to the same townships ("detached
parts").   Speaking generally, we may say that in Buckinghamshire (whatever may be the case in other counties)

whenever Domesday Book (A.D. 1086) gives the name of a township (or as it called it in Latin, a *villa*) that place was a civil parish at the beginning of the nineteenth century. There are exceptions, but they are few. A parish is essentially a place that has a church, but the ecclesiastical unit is intimately linked with the agricultural unit through the payment of tithes; and there can be no doubt that parish-boundaries (excepting those of ancient urban parishes like Stony Stratford) are the agricultural township boundaries of Saxon times.

For example, on the Chiltern hills the parishes are long and narrow and set athwart the escarpment. The churches which mark the sites of the original villages, even if these have disappeared, as in the case of Saunderton and Horsenden, stand along the line of springs where the water flowing underground through the chalk is thrown out by the clay below. The parish areas are so disposed that each has a share of all the different kinds of soil from the chalk downs above to the clay-plain below. Obviously this is an agricultural arrangement, not one for the advantage of the church or the parish priest. Similarly, along the Ouse, parish-boundaries are set at right angles to the river, so that each parish should have access to the rich meadow-land by the water. The same is the case along the Thames, though the boundaries are not so neatly drawn; but the most obvious exception—Taplow —is one that proves the rule, for it occurs just where the bank of the river is lofty and precipitous.

After the time of William the Conqueror, many new settlements were made within the area of the original townships, and these were made into manors (until the act *Quia Emptores* in 1290 forbade this), but in very few

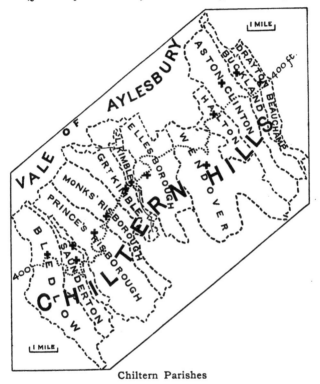

Chiltern Parishes

(The ground to the north-west of the dotted line is less than 400 feet above sea-level, that to the south-east is more than 400. Crosses mark the position of the parish churches. The parish between Saunderton and Princes Risborough is Horsenden)

cases did they have a church built and become parishes. Such settlements are generally distinguished as *hamlets*.

Relics of the days of parish self-government still linger. In some parishes "constables" are still appointed yearly, but they are not members of the police force, and do not seem to have anything to do. In a few villages, such as Thornborough and Dinton, the old parish stocks are still preserved, in which misdemeanants were ordered to be locked.

Village Stocks, Thornborough

But parishes or townships have long ceased to be the social and agricultural units they originally were. A farmer's fields may lie partly in one parish and partly in another; a man's house may lie in one parish and his orchard in another. There is now therefore no meaning in keeping "detached parts" of a parish, and the law has very generally transferred these to the parish surrounding them. It has also, during the last hundred years, made many hamlets into parishes, and altogether made more alterations in the political map of the county than had been made in the previous eight centuries.

The first great change was made after the passing of the New Poor Law in 1834. Under the Old Poor Law of 1601, each parish had been made responsible for the poverty within its borders. For many reasons the parish was too small a unit for this purpose, the county was (in those days when railways were at their birth, and motorcars unthought of) much too large; while the hundreds were almost forgotten. The Poor Law Commissioners therefore set themselves to group the parishes into *unions*, and their main idea in doing so was to combine a number of rural parishes with some urban parish as a centre. In doing this they paid as little attention to county-boundaries as farmers do when they drive to the nearest market-town. Thus in Bucks they took Newport Pagnell, Buckingham, Winslow, Aylesbury, Wycombe, Amersham, and Eton as convenient centres for unions. But in the Wycombe Union they included Stokenchurch and some smaller parishes in Oxfordshire; while a number of Buckinghamshire parishes were included in unions with their centres in other counties—at Brackley, Bicester, Thame, Leighton Buzzard, Berkhampstead, etc. The result of this was eventually to throw all local government areas into great confusion, and though vigorous efforts have been made to simplify the confusion, it has not been possible to do so

without in some cases changing a parish from one county
to another. Thus Stokenchurch, which from time immemorial had lain in Oxfordshire, had to be brought into
Bucks.

Up to the reign of William IV, the parish of Caversfield was a "detached part" of Buckinghamshire lying
within Oxfordshire; while Ackhampstead (in Great
Marlow parish), Boycott (in Stowe parish) and Lilling-
stone Lovell were parts of Oxfordshire lying in Bucks,
and Coleshill (in Amersham) a part of Hertfordshire.
These have now all been merged in their enveloping
county.

One of the most puzzling things about the early
county-boundaries is that they here and there cut a parish
in half. Recent legislation has altered all these cases.
Thus Ibstone and Kingsey parishes lay half in Bucks and
half in Oxfordshire. Now the former is entirely in Bucks
and the latter in Oxfordshire. On the opposite border,
parts of Little Gaddesden, Nettleden, and Puttenham,
which had anciently belonged to Bucks, have been
transferred to Herts, in which the main portion of
each lay.

At the present time the local government of the
county as a whole is in the hands of the County Council,
except for those duties which are undertaken by the
Justices of the Peace. The members of the County
Council are elected by the inhabitants of the county, one
member to a parish or group of parishes in proportion, as
near as may be, to their population. The chief duties
of the County Council are those connected with the

maintenance of main roads and bridges, education (both elementary and higher), public health, the care of the insane, the supervision of weights and measures, the provision of small holdings.   For matters that are purely local,

Old Town Hall, High Wycombe
(*Built* 1757)

or more conveniently dealt with locally, the parishes are grouped into Rural Districts, or if more thickly populated each one may be an Urban District, and for these District

Councillors are elected. High Wycombe is the only town with a Mayor and Corporation.

For the purpose of representation in Parliament, Buckinghamshire is divided into three parts, the North or Buckingham division, the Mid or Aylesbury division, and the South or Wycombe division, each returning one member. Before the Reform Act of 1832, the county returned two members and each of the towns of Buckingham, Aylesbury, Wendover, Amersham, and Wycombe returned a member. Several famous men have sat as members for these boroughs—Edmund Waller for Amersham and Wycombe, Edmund Burke and Canning for Wendover, Wilkes for Aylesbury.

Although Buckingham was the original county-town, its situation so far north must always have been a disadvantage for persons who had to come there from South Bucks on county business. In the reign of Henry VIII, Sir John Baldwin, Chief Justice of the Common Pleas, was an Aylesbury man, and used his influence to have Aylesbury made the county-town, which it now is—the County Assizes being held and County Council business carried on there.

20. Roll of Honour.

Buckinghamshire can hardly claim to be the birthshire of many famous men, but if we include among its heroes those who have chosen it for their home, then they make no mean list. Of poets and statesmen, in particular, Buckinghamshire can claim a goodly share.

The earliest famous person that we read of as born here was St Osyth, daughter of King Frithwald, and grand-daughter of Penda, the Mercian defender of paganism. She was born, some time in the seventh century, at Quarrendon near Aylesbury, and brought up by an aunt at Ellesborough. Though married to Sighere, under-king of the East Saxons, she left him to become a nun, and founded the nunnery of Chich (St Osyth's) in Essex.

We have more certain knowledge of two famous men of the thirteenth century who were born at Wendover. Roger of Wendover became a monk of St Albans and ended as the historiographer of the monastery, that is to say it was his duty to write down the chief events of every year in a chronicle for the information of later generations. In addition to this he made a compilation from the chronicles of his predecessors, and his *Flores Historiarum*, as the whole work is called, is one of the manuscripts from which we learn much of English history. He died in 1236.

Richard of Wendover was a canon of St Paul's in London, but is more famous as a physician. Studying medicine at Paris and Salerno, he became at length physician to Pope Gregory IX, and was the author of a number of medical works. He died in 1252.

Among the religious reformers of the fourteenth century John Wiclif deserves only a brief mention here, as his connection with Bucks is limited to his having been Rector of Ludgershall for six years, while residing at Oxford. Sir John Cheyne, one of the Lollards who took part in Oldcastle's rebellion in 1397, was one of the family whose name is kept by the village of Chenies. From Wendover again there went to London in the fifteenth century one Henry Colet, to become in time Lord Mayor, and to be more famous as the father of Dean Colet, the founder of Westminster School.

St Thomas de Cantilupe (1218–1282), the last Englishman to be canonised, was born at Hambleden. He was a very able administrator, the friend of Simon de Montfort and of King Edward I, Bishop of Hereford and Chancellor of the University of Oxford.

It is in the seventeenth century that Buckinghamshire attains its greatest glory as the home of men and women who took an active part in the great events of the time.

John Hampden, squire of the parishes whose name he bore, described by J. R. Green, the historian, as "a man of consummate ability, of unequalled power of persuasion, of a keen intelligence, ripe learning, and a character singularly pure and loveable," roused the enthusiasm of the Englishmen of his own days and retains the gratitude of those of to-day, for his determined resistance to the attempts of Charles I to tax the people without consulting Parliament. In 1627 he suffered imprisonment for refusing to pay a "forced loan," and in 1637 he was foremost in refusing to pay the "ship-money" illegally demanded by the King. When in 1640, thanks largely to the effects of this refusal, the King was driven to call together the Long Parliament, Hampden was one of his county's representatives. To him was largely due the passing by the Commons of the Grand Remonstrance in 1641.

John Hampden

He was one of the five members whom Charles tried to arrest in the House, and in support of their member five hundred freeholders of Buckinghamshire rode together to Westminster, at the beginning of 1642. When later in the year, the Civil War began, Hampden became an officer in the Parliamentary army, and meeting Prince Rupert, who was making a raid on Wycombe from Oxford, at Chalgrove Field, not far outside the county boundary, he was mortally wounded on June 18th, 1643, and died at Thame six days later.

Among those who fought on the King's side, foremost in Bucks was Sir Edmund Verney of Claydon, Knight-Marshall and Standard-Bearer (1590–1642). "For my part," he said at the opening of the war, "I do not like the quarrel, and do heartily wish that the King would yield and consent to what they desire; so that my conscience is only concerned in honour and gratitude to follow my master. I have eaten his bread and served him near thirty years, and will not do so base a thing as to forsake him; and choose rather to lose my life (which I am sure I shall do) to preserve and defend those things which are against my conscience to preserve and defend." He fell at Edgehill, refusing to save his life by giving up the standard which he carried.

Among other prominent men of this period must be mentioned Bulstrode Whitelocke (1605–1675), of Fawley Court, a successful lawyer and diplomatist, member for Marlow in the Long Parliament, writer and member of many Councils during the Civil War and Commonwealth; Simon Mayne (1612–1661) of Dinton Hall, and

Sir Richard Ingoldsby (d. 1685) of Lenborough, two of the judges who condemned the King to death, the former attainted after the Restoration and dying in the Tower, the latter receiving a full pardon; and Sir Kenelm Digby (1603-1665) of Gayhurst, son of one of the

Sir Kenelm Digby

Gunpowder conspirators, a remarkably versatile man—conversationalist and writer, philosopher, amateur scientist and alchemist, diplomatist and privateersman.

South Bucks was for centuries a centre of religious reform—Lollards in the fourteenth and fifteenth centuries,

Protestant martyrs in the sixteenth century, and Quakers in the seventeenth, were plentiful in and around Amersham and Wycombe. It is of the last that we have the most detailed knowledge, and they may well take a high place in the roll of honour.

Isaac Pennington the younger (1616–1679) was the son of Isaac Pennington, Lord Mayor of London at the opening of the Civil War. The father, like many wealthy Londoners since, had bought an estate in South Bucks, the Grange, Chalfont St Peter; and here, a few years after his marriage, the son settled down in 1658 to a quiet domestic life, which would have been of the happiest kind but for the persecution of the Quakers, which led to Pennington's spending a large part of the ten years that followed the Restoration in Aylesbury gaol.

Mary Proude, afterwards Lady Springett, and finally Mrs Pennington (1624–1682), was a woman of noble mind, great force of character, and the highest practical ability. Married when quite young to Captain Sir William Springett, who was soon after mortally wounded at the siege of Arundel Castle (January, 1644), she made her way to him in the depth of winter and in face of the greatest possible difficulties, though herself in weak health at the time. Later, when married to Isaac Pennington and settled at Chalfont, she won the admiration of the critical Samuel Pepys, who tells us he had "most excellent witty discourse with this lady who is a very fine witty lady, one of the best I ever heard speak, and indifferent handsome" (*Diary*, 8th Oct., 1665). Later, when the

Grange was confiscated on account of Pennington senior's share in the Civil War, the family resided at Berrie House (now called Bury Farm), Amersham; and afterwards Mrs Pennington purchased and rebuilt Woodside House, also near Amersham. During all this time, owing to her husband's frequent imprisonments, the whole manage-

Jordans Meeting House, with the Graves of the
Penns and Penningtons

ment of their affairs as well as the bringing-up of their family devolved upon Mrs Pennington, and only a woman of the highest ability could have succeeded in all as she did. She died in 1682, and both she and her husband are buried at Jordans.

Gulielma Maria Springett, afterwards Mrs William

Penn (1644–1693), was the eldest daughter of Mary
Pennington. Born at Arundel Castle after her father's
death, she grew up to be a woman of great beauty and
high character. Her husband, William Penn (1644–
1718), must be mentioned in any account of Buckinghamshire, though his connection with the county is of
the most indirect kind. His father claimed to be descended
from the Penns of Penn, but the claim is doubtful, and
at most his ancestors must have lived outside the county
for many generations. William Penn himself is buried at
Jordans in South Bucks, and his son Thomas settled at
Stoke Poges at the end of his life and is buried there. But
William Penn's only association with the county during his
lifetime lies in his visits to the Penningtons and his future
wife at Bury Farm, Amersham. Their marriage in 1672
took place at Rickmansworth, and soon after they moved
to Sussex, and Penn sailed across the Atlantic to become
the first "proprietor of Pennsylvania," the state "guarding
in sylvan shades the name of Penn the apostle." It is his
burial with his wife and some of their children and her
mother and step-father in the Quaker burial ground of
Jordans, that draws hundreds of Pennsylvanians yearly in
pilgrimage to South Bucks.

Thomas Ellwood (1639–1713) was born at Crowell
in Oxfordshire, and was with his father a frequent visitor
to the Penningtons at Chalfont, by whose example he
was converted to Quakerism while an undergraduate at
Oxford. He was for a time a constant attendant at the
Meeting-house at Meadle, near Monks Risborough. From
1663 to 1669 he came to reside with the Penningtons,

as tutor to Guli Springett and the younger children, with the former of whom he maintained a life-long friendship. While studying in London shortly before this, he had lodged in Aldersgate near the house of John Milton, and went to read Latin to the blind poet every afternoon.

Milton's Cottage, Chalfont St Giles

To Ellwood we owe Milton's second association with the county, though already in earlier life he had lived at Horton in the extreme south, and had expressed the feelings awakened in him by the country scenes in the poems *Il Penseroso* and *L'Allegro*. When, in June, 1665, the plague became serious in London, Milton asked his friend to find him a country cottage, and Ellwood found the one at Chalfont St Giles, which is preserved as "Milton's Cottage" to this day. Hardly had he hired the cottage, when he found himself imprisoned at Aylesbury for the crime of attending a quaker's funeral at Amersham. Released in August, he soon visited Milton, and was shown the manuscript of *Paradise Lost*. He remarked that the poet had said nothing about Paradise found, and the remark bore fruit, for when he called on Milton in London in the autumn, *Paradise Regained* was shown to him. During the last thirty years of his life, Ellwood lived at Hunger Hill (now called Ongar Hill), between Amersham and Beaconsfield, and he has left a versified direction to find his house from which we learn how much and yet how little the place has changed in two hundred years.

Milton is not the first poet associated with Buckinghamshire, for his contemporary Edmund Waller (1606–1687) was born at Coleshill, where his favourite oak still flourishes (p. 56). He lived most of his life, when not in London, at Beaconsfield. He was a cousin of John Hampden, and related by marriage to Oliver Cromwell. At the age of sixteen he sat in the Commons as member for Amersham, and afterwards represented various other constituencies, including Wycombe. He had no real devotion to the Parliamentary cause, and when sent in 1643 as one of the commissioners to treat with the King at Oxford, he entered into a conspiracy for the surrender of London. For this he was exiled to France, but pardoned in 1651. After the Restoration he sat again in

Thomas Gray

Parliament until his death, and was a favourite with Charles II. His poetry is as different from Milton's as the character of the cavalier from that of the puritan, and much of it is marked by the insincerity of the courtier, though a few poems, such as "Go, lovely rose,"

Stoke Poges Church

deserve not to be forgotten. Waller died at Hall Barn, and is buried in Beaconsfield Church, in the churchyard of which there is a heavy monument with an over-praising inscription.

It was in the eighteenth century that the greatest

poetic glory was shed upon Buckinghamshire. Thomas Gray (1716–1771) was born in London, but his mother was a Buckinghamshire woman, and his early education was undertaken at Burnham by his uncle. At eleven years of age he was sent to Eton, and afterwards went to study law at Cambridge. After his father's death in 1741, his mother and two of his aunts lived at Stoke Poges, and it was while visiting them that he wrote, at intervals, that poem which must endure with the English language—the *Elegy in a Country Churchyard*. Gray saw the poetry and pathos of common, everyday life, and expressed them in simple but most beautiful language. Here too, looking at the scene of his merry school-days, he expressed his sense of life's sadness in the *Ode to the Distant Prospect of Eton College*.

William Cowper (1731–1800) was another poet who, like Gray, wrote about simple things in simple language, at a time when pomposity was the usual character of verse. Born at Berkhampstead, in the part of Hertfordshire that is half embraced by Buckinghamshire, his early life was passed in London. Always a nervous, diffident man, the necessity of passing an examination brought upon him, at the age of 32, the first attack of that madness which re-visited him at intervals during the rest of his life. Soon after his recovery, in 1767, he settled at Olney, with his friend Mrs Unwin, and for a time worked hard helping his friend, the Rev. John Newton, curate in charge of Olney and an earnest evangelical. After a second attack of madness in 1773, he took to a quieter life, amusing himself with gardening, keeping tame hares, and writing poetry. *The Task* was the poem that made his fame, but he is probably better known now from his shorter poems, such as *John Gilpin*, *Boadicea*, and the *Lines on the Receipt of my Mother's Picture*, as well as his hymns, which include "God moves in a mysterious way."

William Cowper

All his published works seem to have been written at Olney. In 1786 he moved to the adjacent village of Weston Underwood, and ten years later to the Norfolk coast, where he died. The house at Olney has now been converted into a museum of Cowper relics.

Among the Buckinghamshire poets must be counted Percy Bysshe Shelley (1792–1822) who lived at Marlow for a short time in 1816–1818, during which time he wrote his *Revolt of Islam*, and published various political tracts under the name of "the Hermit of Marlow."

Cowper's House, Olney

He also exerted himself much in relieving the great distress that then prevailed among his poorer neighbours. The house in which he lived may still be seen at Marlow.

A county that had been so stirred by successive movements of religious reformation could not fail to feel

the influence of Wesley; and under his influence Hannah
Ball (1734–1792), of Wycombe, started in that town one
of the earliest of Sunday Schools, in 1769, twelve years

Shelley's House, Great Marlow

before the date of Robert Raikes's first school in Glou-
cester.

But the sturdy puritanism and quakerism of seventeenth-century Bucks made little stir in the succeeding
century. During that dreary part of English history,

from 1688 to the middle of the nineteenth century, when the English nation, having won the forms of liberty and representative government, allowed the real government of the land to lapse into the hands of a few wealthy families who understood nothing of the real needs of the people—during this time the part played by Buckinghamshire in politics is mainly the part played by Stowe House and the Grenvilles and their friends. As Lord Rosebery has said—" For more than a century ...there had been a Grenville finger in every political pie. During the whole of that time Stowe had been a political fortress or ambuscade, watched vigilantly by every political party; the influence of Stowe had been one which the most powerful minister could not afford to ignore; and the owner of Stowe had been the hereditary chief of a political group." The huge, cold, classic façade of Stowe, completed by Earl Temple in 1760–1780, contrasts with the humble Milton's cottage at Chalfont, as the magnificent political shams of the eighteenth century contrast with the earnest popular movements of the seventeenth.

During this period Buckinghamshire gave two Prime Ministers to the country, both Grenvilles. George Grenville (1712–1770), nicknamed "The Gentle Shepherd," from a mocking quotation with which William Pitt interrupted one of his speeches, was M.P. for Buckingham from 1741 until his death. He began his parliamentary career as member of the group known as the "Boy Patriots," who kept up a persistent attack on Sir Robert's Walpole's administration, and in a few years after Walpole's fall he entered the ministry as Lord of the Admiralty. During the nineteen years in which he was concerned in the administration of the Navy there stands to his credit the ending of the scandal of long arrears in the pay of the seamen, of which Pepys's *Diary* gives us such a vivid idea. In 1763 he became Prime Minister, but his record is one of failure. His Stamp Act initiated that policy towards the colonies which led in the end to the establishment of the United States of America, and his attempts to crush John Wilkes led to the final establishment of the freedom of the press.

George Grenville's youngest son, William Wyndham Grenville, afterwards Baron Grenville (1759–1834), was for a very short time Speaker of the House of Commons; and in 1806, after the death of William Pitt the younger, he became Prime Minister of the "ministry of all the talents," formed of the best men of the three previously existing parties. This ministry was not a success, but it has to its credit the abolition of the slave trade.

Among other eminent members of the same family may be mentioned George Grenville's great-great-grandson, the third and last Duke of Buckingham and Chandos (1823–1889), who was chairman of the London and North-Western Railway from 1853 to 1861, and Governor of Madras at the time of the terrible Indian famine of 1879.

John Wilkes (1727–1797) deserves a place on our roll of honour for his championship of the liberty of the press and liberty of the subject at a time when both were in danger, though in other respects his connection with the

Stowe House

(Seat of the Dukes of Buckingham and long the residence of the exiled Orleans family)

county does it no honour, for he led for some time a wild and dissolute life at Medmenham Abbey. Born in London, where his ancestors had come from Leighton Buzzard some centuries before, he was educated in part at Aylesbury, and was married while still a boy to an Aylesbury heiress, thanks to whose estate he became High Sheriff of the county in 1754 and M.P. for Aylesbury in 1757. His election was secured by means of such open bribery as astonishes us nowadays, but it is only fair to him to say that it was not unusual at that time. After George Grenville's attempt to imprison him for a publication attacking the Government had failed, the House of Commons expelled him; but "Wilkes and Liberty" had become a popular cry, and though he did not again stand for Aylesbury, he fought his way back to Parliament in the end.

Edmund Burke (1729–1797), the famous opponent of the ideas of the French Revolution, was an Irishman by birth, but settled in 1768 at Gregories, Beaconsfield, and was buried in the church there.

The third Prime Minister associated with Buckinghamshire was Benjamin Disraeli, Earl of Beaconsfield (1804–1881). His father, Isaac Disraeli, was a literary man who settled at Bradenham, though the son was born in London. Benjamin Disraeli first distinguished himself, when quite a young man, by a series of novels, which, combined with much that is extravagant and careless in style, show a real insight into the condition of England.

In 1832 and 1834 he unsuccessfully contested High

Hughenden Manor, High Wycombe

Wycombe, and though it was as member for Maidstone that he at last sat in the Commons, he soon bought the estate of Hughenden and lived there for the rest of his life and is there buried. He was the founder of the modern Conservative party in politics. Coming into public life at the time of the Reform Bill of 1832, he recognised that the rule of the old great Whig and Tory families that had lasted since 1688 was now over, and believed in a new kind of Toryism based on the Crown, the Church, and the People. His ideas were regarded as wild and visionary at the time, and he was laughed down when he first spoke in the House. It was more than thirty years before he entered the Cabinet, and forty before he became Prime Minister, so that he had little opportunity to put his youthful ideals into practice, and his policy as a minister is too near our own time to be spoken of impartially.

Few distinguished professional men or men of learning seem to have come from Buckinghamshire. Browne Willis (1682–1760) was the son of a Bletchley landowner, was educated partly at Beachampton, and lived most of his adult life in North Bucks, first at Milton Keynes and afterwards at Whaddon Hall. As a young man he sat for a few years in parliament as member for Buckingham, but his life-work was the study of British antiquities. He was one of the earliest antiquaries to adopt scientific methods, and he travelled all over England and Wales to study the ancient cathedrals.

Sir William Herschel (1738–1822), the great astronomer, was born at Hanover, and came to England as a

musician. He devoted his spare time with enthusiasm
to the study of astronomy, making his own telescopes,
and in 1781 he became suddenly famous by discovering
a new planet, now called Uranus. King George III
appointed him Court astronomer, and to be near Windsor
he settled first at Datchet and afterwards, in 1786, at
Slough. Here he set up his giant telescope, nearly 40
feet long, into which he gazed downwards from a platform 50 feet above the ground, the sky being reflected in
a mirror over four feet in diameter. Here, with the help
of his sister, Caroline Lucretia Herschel (1750–1848), he
worked indefatigably on the surveying of the heavens, and
cataloguing the stars and nebulae. Here he discovered
double stars, one revolving round the other as the earth
revolves round the sun ; here he discovered the direction
in space in which the solar system is moving ; and here
he was enabled to draw far-reaching conclusions as to the
constitution of the Universe. The great French astronomer Arago wrote : "Slough is the place in the world
where most discoveries have been made. The name of
this village will never perish." Sir William Herschel is
buried in Upton church.

Sir John Herschel (1792–1871), the only son of
Sir William, was born at Slough, and received his early
education at Hitcham and Eton. Like his father, he was
a great astronomer, and set up an observatory at the Cape
to do for the stars of the southern hemisphere what his
father had done for those of the northern. He did not
confine himself to astronomy, but made important discoveries in physical optics and in photography. In 1840

he left Buckinghamshire for Kent, and then his father's giant telescope, which had been a landmark for over fifty years, was taken down; but portions of it are still preserved at Observatory House, Slough.

Buckinghamshire has also furnished a President of the

Sir John Herschel

Royal Astronomical Society in Dr John Lee (1783–1866) of Hartwell; and Presidents of the Society of Arts and of the Royal Microscopical Society in the persons of two vicars of Stone—James Booth (1806–1878) and Joseph Bancroft Reade (1801–1870) respectively.

Sir Gilbert Scott (1811–1878) was born at Gawcott near Buckingham, where his father was curate, and was early attracted to the study of architecture. The churches of Hillesden and Maids' Moreton introduced him to Gothic architecture, and Stowe House to the classic style. He grew up to be one of the most famous English architects of the nineteenth century, and some of his work adorns his own county. In his early years of professional work he designed several of the workhouses of the county in Elizabethan style, as at Amersham. Aylesbury and other parish churches were restored under his direction. Perhaps his finest work in the county is Buckingham Church, which he changed from a bare and ugly Georgian structure to a handsome Gothic building.

21. THE CHIEF TOWNS AND VILLAGES OF BUCKINGHAMSHIRE.

(Except in cases where the date 1911 is given, the figures in
brackets after each name refer to the population of the
parish in 1901, an asterisk denoting that there has been a
considerable increase since then. Distances, unless otherwise
stated, are by road. The figures at the end of each para-
graph are references to the pages in the text.)

Amersham (2674*), a picturesque town in the Misbourne
valley, 26 miles by road from London, with station on the Great
Central and Metropolitan Joint Railway. Has the Drake Almshouses (1617), a town-hall on arches over a small market-place
(1682), a quaker meeting-house (seventeenth century), and other
interesting buildings. Near are Raans manor-house, now a
farm-house (Jacobean), and Shardeloes (eighteenth century) in a
fine park. Industries: agricultural; a brewery; lace-making;
chair-making in outlying parts of the parish. (pp. 18, 59, 74, 86,
87, 94, 97, 102, 143, 144, 152, 161, 164, 170, 171, 174, 184, 189.)

Aylesbury (11,048, in 1911), county-town and market-town,
stands on a low hill of Portland stone. The Old Grammar School
is now partly in use as the Museum of the Bucks Archaeological
Society. The King's Head Inn dates from 1444–50. The
London and North-Western station is on the east side of the

town, and the joint Great Central, Great Western, and Metropolitan station on the west. The industries are those of an agricultural market-town, with in addition a condensed-milk factory, and very fine printing and book-binding works. Akeman Street and other ancient roads pass through the town, which was a fortress in very early times. (pp. 4, 12, 36, 42, 49, 67, 90, 95, 100, 101, 102, 104, 137, 139, 140, 146, 147, 148, 151, 153,ª154, 155, 161, 164, 174, 189.)

Aylesbury Market Place

Beaconsfield (2511, in 1911) lies on the main London to Oxford road, here very wide, 23 miles from London. The church is largely of flint, Early English and Decorated, but restored. It contains the remains of Edmund Burke and Edmund Waller, and there is a monument to the latter in the churchyard. The Old Rectory House is a fine early sixteenth-century building, restored. Hall Barn and Wilton Parks and several large beech-woods lie within the parish. Industries: agriculture and chair-making. (pp. 37, 140, 174, 176, 184.)

Bledlow (854) is a village at the foot of the Chiltern Hills, six miles from Thame by road. A tributary of the Thame forms a picturesque gorge, the Lyde, in the Lower Chalk, by the side of which stands the church, of Early English and Decorated styles. There are the remains of a churchyard cross. On the chalk

Beaconsfield: the Oxford Road

downs above the village are many tumuli, and a Cross cut in the turf. Industry: agriculture and paper-making. (pp. 30, 77, 90, 111, 120, 146, 159.)

Bletchley (497), an important junction on the London and North-Western main line, branches diverging to Oxford, Buckingham, and Banbury in one direction, and to Bedford and Cambridge in the other. The town that has grown around the station is almost continuous with Fenny Stratford: the original village lies a mile to the west. The church, in Bletchley Park, is Decorated and Perpendicular, with a Norman doorway, but was restored in the eighteenth century by Browne Willis. (pp. 149, 151, 186.)

Boarstall (151), a small village, remote from everywhere— six miles from Bicester, eight from Thame, nine from Oxford. The moat and tower-gateway are all that remain of Boarstall Castle, which was held for the King in the Civil War and unsuccessfully attacked by Fairfax in 1645. It was formerly in the Forest of Bernwood, and there are woods around still. Remains of ancient cross in churchyard of the modern church. (pp. 104, 137.)

Bradwell (3946). The original village is two miles from Wolverton, but around the station (on the Newport Pagnell branch, L.N.W.R.) has grown up New Bradwell, practically a suburb of Wolverton. The church lies alongside the main line, from which its saddle-back western tower is easily seen. It is Early English and Decorated, with some Perpendicular windows. There was a Benedictine Abbey here, founded about 1155, but no traces of it remain. (pp. 77, 107, 108, 135, 137.)

Brickhill, Great (491), **Bow** (448) and **Little** (278), three villages on or near Watling Street, and two to three miles from Fenny Stratford. Here in the middle ages the Justices in Eyre halted in their journey to try Buckinghamshire cases. The churches are in the main Perpendicular, and more or less modernised. There is a fine view over North Bucks and the adjoining part of Bedfordshire from Bow Brickhill. (pp. 22, 43, 78, 115, 147.)

Brill (1206), a village on the summit of a hill, over 600 feet above sea-level, six miles from Thame, with a fine view over the plain of North Bucks and Oxfordshire. The Saxon and Norman kings had a residence here, for hunting in the Forest of Bernwood. The church has two Norman doorways and a good Early

English window. Industries: agriculture and brick-making. Pottery was made for centuries but is no longer; and ochre was once dug. Portland stone is dug on the hill, and many fossils are to be found. Two old windmills stand on the highest part of the common. (pp. 22, 41, 98, 99, 152.)

Broughton (113), a small village, three miles south of Newport Pagnell, chiefly interesting for its church, which is very early Decorated with Perpendicular windows inserted, and has ancient wall-frescoes of the Last Judgment, St George and the Dragon, etc. The rood-loft stair is preserved, and there are brasses of date 1399 and 1403.

Buckingham (3282, in 1911), the original county-town, chosen as site of a fortress, being at the point where the Ouse passes from the narrow to the broad part of its valley. The original castle has long disappeared, and the modern church stands on the Castle Hill. In the church is preserved a fourteenth-century MS. Bible. The remains of a fourteenth-century market-cross are in the old churchyard. There is a Norman doorway in St John's chantry, lately used as a school. The manor-house has a twisted chimney, almost unique, the only other like it being at Hampton Court. Castle House was rebuilt in 1619–23, on the site of a house in which Catharine of Aragon is said to have received the news of Flodden Field. In the present house, Charles I stayed in June, 1644, and held a council of war, before marching out to the battle of Cropredy Bridge. Buckingham is served by a branch of the Grand Junction Canal, and is the meeting-place of several main roads. The parish is very large, including much agricultural ground, with the hamlets of Bourton, Lenborough, and Gawcott, the last the birthplace of Sir Gilbert Scott. (pp. 2, 3, 12, 23, 36, 104, 126, 137, 140, 143, 149, 153, 154, 161, 164, 189.)

Burnham (3245), a large parish, the main village 3½ miles from Slough and one mile from Burnham Beeches station on the

G. W. Railway. The church has a fine Decorated east window. In the southern part of the parish are some remains of Burnham Abbey (Early English). In the northern part are Burnham Beeches and East Burnham Common, preserved by the Corporation of London for public enjoyment. The beeches are ancient, and having been pollarded at one time have grown to a great thickness without attaining a great height. Within the preserved area is an ancient camp. Industries: agriculture and brickmaking. (pp. 37, 54, 55, 105, 115, 135, 137, 177.)

Chalfont St Giles (1362), a large village in the Misbourne valley, near the main Aylesbury road, 22½ miles from London, and three miles from each of the stations, Chalfont Road (Great Central and Metropolitan) and Gerrard's Cross (Great Western and Great Central). The church has a Norman foundation with Decorated and Perpendicular superstructure, and contains a wall-painting of the daughter of Herodias dancing, and a palimpsest and other brasses. It is approached through a picturesque archway. Milton's Cottage is preserved: in it the poet lived during the plague of 1665. Two miles from Chalfont, on the road to Beaconsfield, is Jordans meeting-house, built by Isaac and Mary Pennington, 1687, in the graveyard of which are the tombs of the Penningtons and Penns, including William Penn of Pennsylvania. Industry: agriculture. (pp. 78, 97, 98, 108, 170, 171, 172, 173.)

Cheddington (578), a small village, with a station on the London and North-Western main line, the junction for Aylesbury, from which it is distant seven miles by rail. There is a small Perpendicular church. On the slopes of South End and West End hills (outlying masses of chalk at some distance from the Chiltern escarpment) are a very fine series of lynchets, or ancient cultivation-terraces, which are plainly seen from the main line trains. Industry: agriculture, chiefly dairy-farming. (p. 112.)

Chesham

Chenies (324), a picturesque village in the Chess valley, 4½ miles from Chesham. The church is mainly Perpendicular, but its chief interest lies in the private chapel of the Russells, which contains many tombs of the family from 1556 onwards. The manor-house was rebuilt in the time of Henry VIII, around a quadrangle, but only the south side now remains and it is remarkable for having no windows on the south face. (pp. 78, 140, 166.)

Chepping Wycombe, see **High Wycombe**.

Chesham (8204, in 1911), a growing town in the Chess valley, five miles S.W. of Berkhampstead. The church, Decorated and Perpendicular, was restored by Sir Gilbert Scott. Large blocks of Eocene conglomerate serve as foundation-stones. On the valley-side above the railway-station are lynchets, known as the Balks. Industries: boot-making, brush-making, and the making of wooden-ware. It is also a market-town, and at Cowcroft in the western part of the town there are large brickfields in which chalk and sand are also dug. (pp. 27, 28, 47, 78, 81, 87, 89, 90, 97, 102, 108, 112, 120, 152.)

Chesham Bois (767*), on the high ground between the Chess and Misbourne valleys—1½ miles from Chesham, and one mile from Amersham station, its nearness to which has made it a residential suburb for Londoners. There is a pretty common, and the church (Early English and Decorated) contains some fourteenth-century glass.

Chetwode (157), a very small village, 5½ miles from Buckingham. The church is Early English, with a beautiful south window with three lancet lights containing thirteenth-century glass. (pp. 127, 129, 135.)

Chilton (285), a small village, on a ridge, four miles from Thame, three miles from Wotton station (G.C.R.). The church has very good Early English, Decorated, and Perpendicular work. (p. 42.)

Cholesbury (107), a very small parish four miles from Chesham, celebrated as the one in which, under the old Poor Law, the rates in 1832 absorbed the whole produce of the place. There is a large oblong camp, within which stands the church, mainly Decorated but with an Early English south doorway. (p. 115.)

Claydon, East (336), **Middle** (231), **Steeple** (721), three villages, with the hamlet of Botolph Claydon, on the clay-plain of North Bucks, three to five miles from Winslow, and having five railway stations within a radius of two miles (L.N.W.R., G.C.R., and Metropolitan). Almost the only villages in the county with free public libraries. Claydon House, in a park, is partly Tudor, with an eighteenth-century front. Industries: agriculture; large brickfields by Calvert station. (pp. 17, 36, 78, 143, 144, 168.)

Clifton Reynes (122), a village in the Ouse valley, one mile by footpath, 3½ miles by road from Olney. Church, early Decorated, with good clustered piers and two fine altar-tombs.

Coleshill (535), two miles from Amersham, a village on a hill with fine views, eastwards to Harrow, southwards over the Thames valley to the Surrey Downs, Windsor Castle, and Hind Head. It was formerly a detached part of Hertfordshire. The birthplace of the poet Waller, it still has Waller's oak. (pp. 56, 78, 162, 174.)

Colnbrook, a village on the Bath road where it enters the county, included in the parish of Horton (which see). It is 17 miles from London, and three from Slough. (p. 149.)

Creslow (5), an almost uninhabited parish, with one house, an Elizabethan manor-house with out-buildings in which there is a Norman doorway.   In Tudor times the cattle from the Creslow pastures supplied beef to the royal table.   (pp. 71, 108, 140.)

Cublington (215), seven miles N.E. of Aylesbury, has a large artificial mound which may be a tumulus, and a small Perpendicular church.   (p. 118.)

Cuddington (455), a village on high ground above the river Thame, six miles S.W. of Aylesbury.   The church is of various styles from Norman to Perpendicular, with a good Early English doorway.   (pp. 108, 129.)

Datchet (1834), a residential town by the Thames, 1½ miles from Windsor.   (pp. 37, 187.)

Denham (1146), a picturesque village on the Misbourne near its confluence with the Colne, two miles from Uxbridge. There is much fishing in the two rivers.   (p. 152.)

Dinton (663), a village 4½ miles S.W. of Aylesbury.   Anglo-Saxon burials have been found in the parish.   The church has a very fine Norman doorway with carved tympanum, and the steps of the churchyard cross and the village stocks are preserved. Dinton Hall, partly Jacobean, but with many later alterations, was the residence of Simon Mayne, and Oliver Cromwell slept there on 14th June, 1645, after the battle of Naseby and left behind his sword, which is still kept there.   John Bigg, who was clerk to Simon Mayne, is traditionally said to have been the executioner of Charles I.   In his later life he was known as the Dinton hermit, living in a cave in the village.   (pp. 51, 117, 126, 137, 160, 168.)

Dorton (140), a very small village, one mile from Wotton station (G.C.R.).   A mineral spring, issuing from the Kimmeridge clay here, was at one time in some repute as a spa.

Edlesborough (1030), a large village on the Bedfordshire border, three miles from Dunstable.   (p. 146.)

Ellesborough (577), on the Upper Icknield Way, 4½ miles from Aylesbury, and one mile from Little Kimble station, at the foot of the Chiltern Hills, which here attain their greatest height at Coombe Hill (852 ft.), at the point where the monument stands to the memory of Buckinghamshire men who fell in the Boer War. The church (Perpendicular) is built on an ancient tumulus. Pit-dwellings, probably of Neolithic age, have been found in the neighbourhood. Chequers Court, built at the end of the fifteenth

Head Master's Room, Eton College

century, with eighteenth-century additions and alterations, contains various relics of Oliver Cromwell, including the sword which he used at Marston Moor. The box-tree grows abundantly on the chalk soil of this neighbourhood. (pp. 51, 107, 108, 114, 159, 165.)

Emberton (120), 1½ miles from Olney. The church has much good work of the Decorated style, especially the east window. (pp. 130, 131.)

Eton (3192, in 1911), an urban district on the left bank of the Thames, opposite Windsor. Famous for Eton College, one of the great public schools, founded by Henry VI in 1440. The chapel is a fine example of the Perpendicular style. The main buildings date from about 1517. (pp. 92, 131, 134, 140, 141, 161, 177, 187.)

Farnham Royal (1162), a large parish with scattered population; the village 2½ miles from Slough, and near Burnham Beeches.

Fenny Stratford (5171, in 1911), a town on the Watling Street where it crosses the river Ouzel; also on the Grand Junction Canal, and one mile from Bletchley Junction. The church, an uninteresting seventeenth-century building, contains Browne Willis's tomb. (pp. 12, 24, 78, 108, 146, 147.)

Fingest (367), a village hidden among the Chiltern woodlands, 6½ miles N.W. of Marlow. The church has a Norman tower and nave, with additions in later styles. Industries: agriculture and chair-making.

Fleet Marston (53), a depopulated village, three miles N.W. of Aylesbury along Akeman Street. The small Decorated church stands alone. Industry: agriculture, chiefly grazing. (pp. 16, 71, 78, 108, 121.)

Foscott (46), a small village two miles N.E. of Buckingham where remains of a Roman villa have been found. (pp. 77, 114.)

Gayhurst (133), in the Ouse valley, three miles north of Newport Pagnell. In the park stands the fine Elizabethan mansion, birthplace of Sir Kenelm Digby in 1603. (pp. 140, 142, 169.)

Gerrard's Cross (552*), a modern parish, made out of parts of five old parishes (Chalfont St Peter, Fulmer, Iver, Langley Marish, and Upton-cum-Chalvey), with a station on the

Great Western and Great Central Joint Railway. A favourite residential district for Londoners. There is a large common, and in Bulstrode Park are the remains of a very large camp.

Grendon Underwood (326), near Akeman Street, 10 miles N.W. of Aylesbury, and two from Akeman Street station. At the Ship Inn, Shakespeare is said to have slept. (pp. 78, 121, 147, 152.)

Haddenham (1223), a large village, three miles N.E. of Thame. The church has a transition Norman chancel arch, the rest being in various later styles. Industry: agriculture. Portland stone is dug in many small pits, but for local use only. (pp. 42, 99, 106, 152.)

Hampden, Great and **Little** (384), two villages among the Chiltern woodlands, about 7½ miles from Aylesbury and three to four miles from Great Missenden station. Little Hampden church has thirteenth-century wall-paintings. Hampden House is mainly of eighteenth-century date, but incorporates parts of older houses on the site dating back to the fourteenth century. It was visited by Queen Elizabeth, on which occasion it is said that the fine avenue of beech and Spanish chestnut was made by cutting through the woods. Considerable remains of Grim's Dyke and other earthworks are to be seen in the woods. (pp. 37, 78, 108, 140, 166.)

Hanslope (1424), a village six miles N.W. of Newport Pagnell. The church is Perpendicular, with some remains of Norman and Early English work, and is noted for its very fine spire, which was destroyed by lightning in 1804 and restored. It forms a conspicuous landmark. (p. 137.)

Hartwell (118), two miles from Aylesbury. In the park stands Hartwell House, built early in the seventeenth century, but largely altered in the eighteenth, famous as the residence of the exiled King Louis XVIII and his court, and later of Dr John

Lee, F.R.S., the astronomer. Industries: agriculture and brick-making. The clay dug for bricks is famous for the abundant fossils it contains, some being of species not found elsewhere in England, so that it has received the distinctive name of Hartwell Clay. (pp. 77, 105, 107, 117, 188.)

High Wycombe (20,390), the largest town in Bucks, a corporate borough, in the valley of the river Wye or Wyck,

Wycombe Abbey School

on the main line of the Great Western and Great Central Joint Railways, and on one of the main motoring roads from London to Oxford. Industries: chair- and furniture-making, with a considerable home and export trade, and paper-making. Wycombe Abbey School ("the Eton for girls") ranks among the foremost of girls' public schools, and in the town there are also a Grammar School, a County High School, and Godstow Preparatory School, besides minor schools. The town is very ancient, important Roman remains having been found indicating permanent civilised settlement in the Roman period. An ancient British trackway runs up the hillside alongside the Amersham road, but is being gradually obliterated. Ruins of a late Norman hospice are standing in the grounds of the new Grammar School. The church is Early English, with many additions and considerably restored. (pp. 6, 27, 67, 85, 87, 90, 94, 96, 97, 102, 108, 114, 118, 128, 129, 137, 151, 152, 161, 163, 164, 168, 170, 180, 186.)

Hillesden (181), 2½ miles by footpath from Claydon station (L.N.W.R.). The church is a very beautiful example of late Perpendicular style. There formerly stood by it the mansion which Sir Alexander Denton fortified and held for the King in the Civil War, and which was besieged and taken by Oliver Cromwell in February and March, 1643. Bullets then fired were to be seen in the church door until quite recently. The churchyard cross is the best preserved in the county. (pp. 16, 104, 131, 133, 189.)

Hitcham (553), a long and narrow parish, lying between Taplow and Burnham, and touching Taplow station on the Great Western Railway. The church contains a Norman chancel arch, and a very perfect Decorated chancel, in which much contemporary stained glass and encaustic tile-work remains. (pp. 107, 121, 187.)

Hoggeston (129), a small agricultural village, three miles from Winslow. There are remains of the ditch by which the village site was surrounded for protection in early days—an oblong site, nearly twice as long as broad, having the church in its north-east corner. There was formerly much straw-plaiting done here. (p. 87.)

Hogshaw-with-Fulbrook (56), one of the "deserted villages" of the North Bucks plain, and one of which the exact history of its depopulation is known. The greater part of the parish (77 per cent., according to the statement of acreage, which may not be exact) was enclosed for pasture in March, 1489, including both the village of Hogshaw and hamlet of Fulbrook; 11 houses were pulled down and 60 persons evicted. Hogshaw church was partly destroyed in the Civil War, and finally pulled down in 1730. The font still exists, used to grow flowers in, in front of the door of Fulbrook House, an Elizabethan building now a farm-house. (pp. 16, 71.)

Horton (834), a village in the Colne valley, four miles from Slough, 1½ miles from Wraysbury station (L.S.W.R.), the home of Milton in early manhood, where he wrote *Lycidas*, *Il Penseroso*, and *L'Allegro*. The church is Early English and Perpendicular, with a Norman doorway. Industry: paper-making. (pp. 90, 126.)

Hughenden (1728), 1½ miles from High Wycombe. The house, in the park, was the residence of Benjamin Disraeli, Earl of Beaconsfield, twice Prime Minister. His tomb adjoins the church, which lies in the park and is of mixed styles with a plain Norman doorway. (pp. 47, 96, 185, 186.)

Ibstone (236), a small village among the Chiltern woodlands, eight miles from Marlow, 8½ from Wycombe and 4½ from Aston Rowant station. From at least the eleventh to the nineteenth century the parish was half in Oxfordshire. The church is partly Norman, and there is a very ancient yew-tree in the churchyard. (pp. 56, 162.)

Ickford (319), a village in the Thame valley, 4½ miles from Thame, 1½ miles from Tiddington station. The church is mainly good Early English, but has two Norman windows, and others of Decorated, Transitional, and Perpendicular style.

Iver (2623), a large village, in the Colne valley, 2½ miles from Uxbridge and two miles from each of two stations—Langley on the main line and Cowley on the Uxbridge branch of the Great Western Railway. The church dates probably from the

Saxon period, being then aisleless. Remains of the original Saxon windows are seen, blocked up and cut through in making the aisle in Norman times. The chancel arch is Early English. Tradition reports that the church was burned by the Danes. Industries: brick-making and market-gardening. (p. 124.)

Ivinghoe (818), an agricultural village at the foot of the Chilterns, and on the Lower Icknield Way, nine miles east of Aylesbury. On Beacon Hill there is a tumulus. The church is cruciform, with central tower, Decorated. The name occurring in an old rhyme—

> "Tring, Wing, and Ivinghoe
> For striking of a blow
> Hampden did forgo,
> And glad that he escaped so,"—

suggested to Sir Walter Scott the title of his novel, and name of its hero, *Ivanhoe*; but the scene of the novel is not set in Bucks at all. The rhyme in question does not appear to have any historical basis. (pp. 120, 135, 146.)

Kimble, Great (345) and **Little** (158), two agricultural villages at the foot of the Chilterns, 4½ miles south by rail and six by road from Aylesbury, with Little Kimble station on the Great Western and Great Central Joint Railway. The name Kimble derives from Cymbeline, and a fortified mount, though probably Neolithic in age, is traditionally ascribed to Cymbeline and may well have been occupied by him. There is another probably Neolithic hill-fortress on Pulpit Hill. Great Kimble village is on the Upper, and Little Kimble on the Lower Icknield Way, but both churches (Decorated) are on the Upper. To the north-west of Great Kimble church is a large barn of fifteenth-century date. Industry: agriculture, duck-farming. (pp. 14, 73, 78, 83, 101, 146, 159.)

Langley Marish (2801), two miles east of Slough, with a station on the main line of the Great Western Railway. The church is Decorated and Perpendicular, and there is a very large yew in the churchyard. (p. 56.)

Lathbury (159), in the Ouse valley, one mile from Newport Pagnell. The church, originally aisleless, built in very early Norman times, had aisles added later in the same period, and the blocked up tops of the original windows are still seen. There are also Early English and Decorated portions, and a series of wall-paintings dating from the fourteenth to the seventeenth centuries. (p. 124.)

Latimer (583), a picturesque village in the Chess valley, 3½ miles from Chesham, one mile from Chalfont Road station. Latimer House is a fine Elizabethan mansion. (pp. 78, 114, 142.)

Lavendon (674), three miles from Olney. The church has a possibly Saxon tower and the nave arcading is transition Norman to Early English. In the parish are earthworks, surrounding the site of Lavendon Castle; also the site of Lavendon Abbey. (pp. 123, 137.)

Linslade (2262, in 1911), at one time a depopulated village, has now become a suburb of Leighton Buzzard in Bedfordshire, and contains the Leighton station of the London and North-Western Railway. The church is a mile and a half distant, small, Decorated and Perpendicular, with remains of the church-yard cross.

Long Crendon (1075), a large village on a ridge extending from the Thame valley to Brill. It is two miles north of Thame. A very ancient road, the Angle Way, runs through it from N.E. to S.W., and alongside this is the site of a Roman cemetery. The church is Early English to Perpendicular, and from the tower there is a very fine view. Near the church is the fifteenth-century Court House, lately restored by the National Trust. In

the parish, near the river Thame, are the remains of Notley
Abbey, Early English work incorporated into a farmhouse. For
several centuries needle-making was an important industry, but it
became extinct about 50 years ago. Portland stone and sand are
dug, chiefly for local use. (pp. 42, 95, 115, 138, 146.)

Great Marlow, from the Thames

Ludgershall (325), a picturesque agricultural village, six
miles S.E. of Bicester, with a station on the Great Western
Railway (Birmingham line). (pp. 152, 165.)

Maids' Moreton (425), 1½ miles N.E. of Buckingham on
the Towcester road. The church was entirely rebuilt in 1450
by two unmarried sisters, whence the prefix to the place-name.

It is Perpendicular, with hanging tracery on the tower, and fan-tracery in the north porch and over the west doorway. The rood-screen is well preserved. (pp. 131, 132, 189.)

Marlow, Great (4683, in 1911), a town on the Thames, here crossed by a suspension bridge. There is a Grammar

Grammar School, Great Marlow

School in the town and the house occupied by the poet Shelley still stands. There is also a large brewery, and the place is a favourite summer boating resort for Londoners. (pp. 12, 22, 78, 95, 179.)

D. B.   14

Medmenham (387), a village on the north bank of the Thames, three miles from Marlow, four from Henley. The ruins of Medmenham Abbey, partly incorporated into a dwelling-house, are seen from the river. The church is mainly Decorated, but has a Norman doorway. (pp. 135, 136, 184.)

Mentmore (289), four miles from Leighton Buzzard, two from Cheddington station (L.N.W.R.), with a large modern mansion and park.

Milton Keynes (219), a small village, four miles from Fenny Stratford and four from Newport Pagnell. The church is mainly Decorated, with very beautiful work, but a narrow transition Norman chancel arch remains from the older church. (pp. 129, 186.)

Missenden, Great (2166*), a small but growing town in the Misbourne valley, four miles from Chesham, now a favourite residential place for Londoners. The church is Early English to Perpendicular. There was once an important abbey here. (pp. 78, 135.)

Missenden, Little (1112), a pretty village in the Misbourne valley, 2½ miles from Amersham and the same distance from Great Missenden. The church is an interesting example of very plain Norman work, the nave and chancel having been built early in the twelfth century, and the aisles added in the latter part of the same century. Industry: agriculture. There are many beech woods in the parish, and in the hamlet of Holmer Green, chairs and tennis-rackets are made. (pp. 78, 126.)

Monks Risborough (714), a village at the foot of the Chilterns, 1½ miles from Princes Risborough station, 5½ from Aylesbury. On the escarpment-face of the Chilterns is cut the Whiteleaf Cross, a landmark for many miles over the plain. The township from very early times belonged to the Archbishop of Canterbury, hence the prefix to the place-name. The church is good late Early English and Decorated. Industry: agriculture and duck-farming. (pp. 78, 82, 110, 146, 159, 172.)

Newport Pagnell (4239, in 1911), a market-town on the Ouse, and terminus of a branch of the London and North-Western Railway. It was an important post of the Parliamentary forces during the Civil War. The church is large, of Decorated and Perpendicular styles. There are carriage and motor-building works, and a brewery. (pp. 24, 92, 104, 117, 118, 135, 154, 161.)

North Crawley (541), a small village on the Bedfordshire border, four miles east of Newport Pagnell. The church is Decorated and Perpendicular, and has a fine Decorated rood-screen, with painted figures.

North Marston (524), three miles from Winslow and from Winslow Road or Grandborough Road stations. A famous place of pilgrimage in the fourteenth and fifteenth centuries, until the shrine of Master John Schorne was removed to Windsor. The church is a very fine one, mainly Decorated, but with some Early English and a good Perpendicular chancel, with carved miserere seats. (p. 78.)

Olney (2684), a market-town on the Ouse, at the extreme north of the county and within the Northamptonshire boot-making area. There is a large and beautiful Decorated church. The Ouse is crossed by a bridge of five arches. Famous as the home of the poet Cowper for the years during which most of his poems were written: his house is now a Cowper museum. (pp. 12, 24, 25, 89, 151, 177, 178, 179.)

Penn (1030*), a village on high ground, three miles from High Wycombe, 2½ from Beaconsfield station. In the neighbourhood are many beech and fir-woods, and the view from Beacon Hill extends over parts of six or seven counties. The church is Early English, but much rebuilt.   It contains interesting brasses.   Industries: agriculture and chair-making.   It is also a residential district for Londoners.   (pp. 22, 37, 44.)

Preston Bissett (290), a small village four miles by road from Buckingham.   The church is a beautiful example of Decorated work.   (p. 131.)

Market House, Princes Risborough

Princes Risborough (2189), a small town at the foot of the Chilterns, near the Upper Ickneild Way, with an important junction (one mile away) on the Great Western and Great Central Railway, four lines here diverging.   Situated in one of the gaps in the Chilterns, it was anciently of much importance, being a royal manor and the meeting-place of one Witenagemot (A.D. 884); but its importance has long been overshadowed by that of

Aylesbury, six miles distant. Most of the population now lives in hamlets both on the hills and in the plain. The church contains some beautiful Early English work. Industries: agriculture, chair-making, bead and sequin-work. (pp. 43, 62, 78, 100, 107, 146, 152, 159.)

Quainton (838), a picturesque village, beautifully situated at the foot of a hill reaching 610 feet above the sea, eight miles from Aylesbury. The church is Decorated and Perpendicular, with a rood-loft staircase and interesting brasses. There are almshouses of date 1687, and near the village is a moated house, the only inhabited one in the county. On the village green is the stump of the ancient market-cross. Portland stone is dug on Quainton Hill. (pp. 7, 22, 23, 36, 42, 108, 119, 120, 140, 143, 152.)

Quarrendon (65), once an important village, the birthplace of St Osyth, depopulated at some time unknown. There was a mansion here as late as the time of Queen Elizabeth, who visited it. Now only the foundations remain, and the church adjoining it, once a beautiful Decorated building, is in ruins. (pp. 16, 71, 165.)

Radclive (295), a village in the Ouse valley, two miles from Buckingham. Church very early Early English in the main.

Saunderton (368), one of the very long and narrow Chiltern parishes. Saunderton station (Great Western and Great Central Joint Railway) is near the large workhouse of the Wycombe Union. The church (Decorated and modern) is three miles away, and less than a mile from Princes Risborough station. There is no village. There are tumuli on the downs. (pp. 120, 158, 159.)

Shenley, two villages—**Church End** (166) and **Brook End** (186), about four miles N.E. from Fenny Stratford and the same from Stony Stratford. There is a camp and a moat at Church End. (pp. 114, 115.)

Simpson (731), a village in the Ouzel valley, 1½ miles north of Fenny Stratford. The church is cruciform, with central tower, and without aisles; mainly of the Decorated period.

Slough (14,985, in 1911), a modern town on the Bath Road, and Great Western Railway. There are brickfields, engineering works, embrocation works, and seed-testing grounds. At Salt Hill is a tumulus, and the famous old coaching inn, the Windmill. Slough was famous for many years for Sir William Herschel's giant telescope. (pp. 37, 62, 64, 95, 96, 98, 99, 118, 149, 151, 153, 187, 188.)

Soulbury (550), a village on the side of the Ouzel valley, 2½ miles from Leighton station (London and North-Western Railway). A large erratic of Carboniferous limestone from the Glacial Drift stands in the roadway. (p. 48.)

Stewkley (1159), a large village, five miles N.W. from Leighton Buzzard. The church is an exceptionally perfect example of late Norman date. There are brickfields in the Kimmeridge Clay, in which very large septaria are found, and stone-pits in the northernmost outcrop of Portland stone in England. Straw-plaiting was formerly carried on here. (pp. 87, 124, 125, 126.)

Stoke Mandeville (282), a village 2½ miles S.E. from Aylesbury. The old church, now falling into ruins, is mainly Decorated, with some earlier and later work. (p. 78.)

Stokenchurch (1600), after belonging to Oxfordshire from time immemorial, has been given to Bucks in recent years. It stands high on the Chiltern plateau, over 700 feet above sea-level, on one of the main London and Oxford roads, seven miles west of West Wycombe and eight from Marlow. There are beech-woods all around and chair-making is the chief industry. (pp. 161, 162.)

Stoke Poges (1398), a village two miles north of Slough. The church has a Norman chancel arch, Early English and Decorated aisle windows, and a Perpendicular east window. The churchyard is famous as the scene of Gray's *Elegy*. There is a very heavy and inappropriate monument to the poet. (pp. 78, 172, 176, 177.)

Stone (1393), a village on the ancient road from Thame to Aylesbury, 2½ miles S.W. of the latter town. The County Lunatic Asylum is here. There are remains of Roman and Saxon burial-places, a stone-quarry (Portland and Purbeck beds) and sand-pits (Lower Greensand). The church is mainly Early English, with some Norman work and later additions. The steps of the churchyard cross remain. (pp. 22, 42, 115, 117, 188.)

Stony Stratford (2353), a market-town at the point where Watling Street crosses the Ouse, also on the Grand Junction Canal, and 2½ miles (by steam-tram) from Wolverton station. Remarkable as an inland ship-building town, making steam launches for canals and even exporting them abroad. (pp. 12, 24, 78, 92, 114, 147, 158.)

Stowe (246), 2½ miles N.W. of Buckingham. The village was bodily removed to make room for the great mansion, which was completed by Earl Temple with Adam as architect in 1760–80. Stowe House stands in a large park, and is approached from Buckingham by an avenue a mile and a half long. The Orleans royal family of France occupied the house for many years. (pp. 36, 54, 105, 144, 148, 149, 181, 183, 189.)

Taplow (1056), two miles east of Maidenhead, with a station on the Great Western Railway. A favourite summer resort for boating on the Thames, which here flows below steep banks beautifully wooded. From a tumulus on the site of the ancient church many Saxon antiquities have been obtained. (pp. 30, 106, 107, 109, 117, 143, 158.)

Thornborough (481), a village three miles east of Buckingham. The stocks are preserved. The church is Decorated with Perpendicular alterations. On the road to Buckingham are two large tumuli. (pp. 108, 118, 160.)

Turville (371), a village in the Chiltern woodlands, 7½ miles west of Wycombe. The church is Decorated and Perpendicular, but has a doorway showing transitional features from Norman to Early English.

Twyford (340), five miles S.S.W. of Buckingham. The church is mainly Early English, with a Norman doorway. (p. 77.)

Upton-cum-Chalvey (8), a parish reduced almost to nothing by the extension of Slough at its expense. The church has a Norman chancel arch, doorway and window, and has been claimed as the scene of Gray's *Elegy*. (pp. 126, 187.)

Waddesdon (1523), a large village on Akeman Street, 5½ miles by road N.W. of Aylesbury. There is a large park with a mansion on a hill-top, forming a conspicuous landmark. (pp. 77, 108, 122, 147.)

Water Stratford (113), in the Ouse valley, four miles west of Buckingham. The church has two Norman doorways. (pp. 9, 24, 78, 121, 126, 148.)

Wavendon (1583), one mile from Woburn Sands station on the Bedford branch of the London and North-Western Railway, and 3½ miles from Fenny Stratford. The Lower Greensand here forms a conspicuous ridge of wooded hills, and there is an ancient hill-top fortress called "Danesborough." There are large brickfields near the station. (pp. 53, 54.)

Wendover (2009), a large village at a gap in the Chilterns on the Upper Icknield Way, five miles from Aylesbury, with a station on the Metropolitan and Great Central Joint Railway, and the terminus of a branch of the Grand Junction Canal. The church is Decorated, but restored. The hills in the neighbourhood are a favourite holiday resort. (pp. 13, 22, 37, 85, 100, 108, 146, 153, 164, 165, 166.)

Weston Turville (720), a village at the foot of the Chilterns, three miles S.E. of Aylesbury on the Lower Icknield Way, with a large reservoir for the Wendover Canal. Duck-farming is carried on. (pp. 82, 87, 115.)

Weston Underwood (275), two miles S.W. of Olney. Cowper lived here from 1786 to 1796. (pp. 78, 178.)

West Wycombe (3466), two miles from High Wycombe on the Oxford road. On a high hill (500 feet) above the village between two parallel valleys, stands an ancient earthwork-fortress, containing within it the church, and the Dashwood mausoleum (eighteenth century). The site of Desborough Castle lies within the parish, but near High Wycombe. Industries: chair-making and agriculture.

Whaddon (321), five miles south of Stony Stratford. The church is of mixed styles from transition Norman to Perpendicular. There are some Norman remains of Snelshall Priory. (pp. 108, 109, 186.)

Whitchurch (619), a picturesque village, five miles north from Aylesbury on the Buckingham road. Traces of the site of Bolbec Castle may be seen. The church is of various styles, and has a fine Early English tower-doorway. A hiring fair was formerly held here in October. (pp. 69, 70, 137.)

Willen (91), a small village in the Ouzel valley, 1½ miles south of Newport Pagnell. The church was built by Sir Christopher Wren for the vicar, his old schoolmaster, Dr Busby.

It was originally a plain rectangular building, but an apse has now been added. (p. 133.)

Wing (1740), 2½ miles from Leighton station (L.N.W.R.). The church has one of the six Saxon crypts of England, which has above it a Saxon apse. There are two rood-lofts and the chancel-arch has a rood-screen. In the churchyard are the remains of a cross. There are tumuli in the neighbourhood. Industries: agriculture and brick-making. (pp. 46, 122, 123.)

Wingrave (827), six miles N.E. of Aylesbury. The church is Decorated and Perpendicular, but on the north side of the chancel there is arcading of transition Norman style. (p. 129.)

Winslow (1703), a small town, 10 miles from Aylesbury and 6½ from Buckingham, on the main road between those places. From the foundation of St Albans Abbey by Offa, in 796, Winslow belonged to the abbey. The church is mainly Perpendicular, with some Decorated windows. (pp. 100, 114, 146, 161.)

Wolverton (5323), a town of modern growth, two miles N.E. of Stony Stratford, dependent on the carriage-building shops of the London and North-Western Railway. There is also an envelope factory. (pp. 6, 62, 67, 69, 70, 90, 91, 92, 149, 150, 151, 153.)

Wooburn (3328), in the Wye or Wyck valley, five miles S.E. from Wycombe. There are large paper-mills. The church is Decorated and Perpendicular. (p. 90.)

Wotton Underwood (235), a very small village, eight miles W.N.W. of Aylesbury, taking its name from its proximity to Bernwood Forest. There are still many oak-woods in the neighbourhood. Wotton House, rebuilt by the Duke of Buckingham and Chandos, stands in a large park. (p. 78.)

Wraysbury or **Wyrardisbury** (779), a village on the Thames flats opposite Old Windsor, two miles from Staines. The church is mainly Early English; and there are some traces of the Benedictine nunnery of Ankerwyke by the Thames. Magna Carta Island is in this parish. There are paper-mills on the Colne, which runs through part of the parish. (pp. 29, 90, 102, 103, 121.)

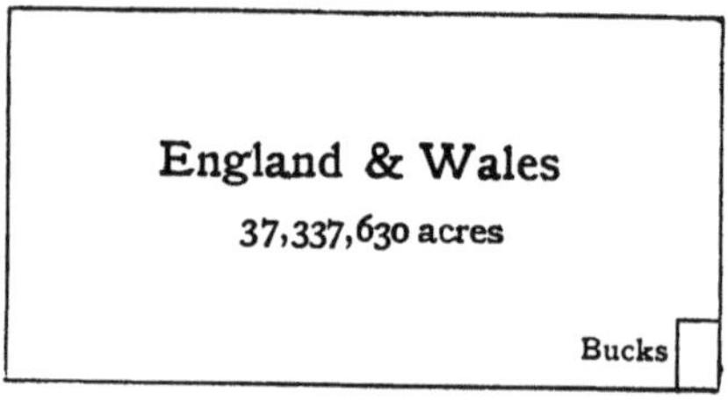

England & Wales
37,337,630 acres

Bucks

Fig. 1. Area of Buckinghamshire (479,359 acres)
compared with that of England and Wales

England & Wales
36,075,269

Bucks

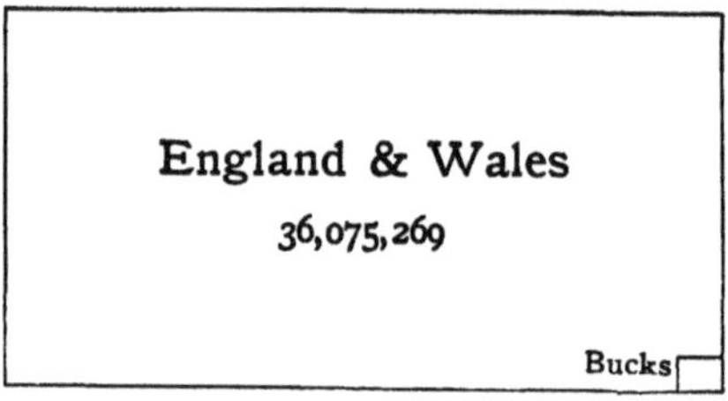

Fig. 2. Population of Buckinghamshire (219,583) compared
with that of England and Wales in 1911

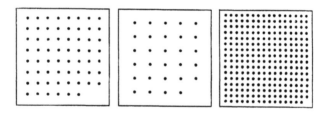

England and Wales 618   Buckinghamshire 293   Lancashire 2550

Fig. 3. Comparative density of Population to the
square mile in 1911

(Each dot represents 10 persons)

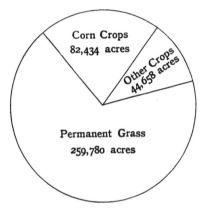

Fig. 4. Proportionate areas of Corn and Other Crops
in Buckinghamshire in 1910

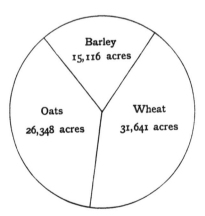

Fig. 5. Proportionate area of Chief Cereals in
Buckinghamshire in 1910

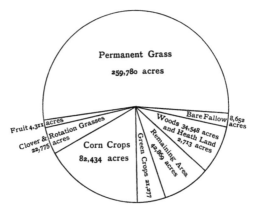

Fig. 6. Proportionate areas of Land in Buckinghamshire
in 1910

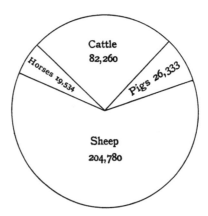

Fig. 7.   Proportionate numbers of Horses, Cattle, Sheep,
and Pigs in Buckinghamshire in 1910

Milton Keynes UK
Ingram Content Group UK Ltd.
UKHW041520181024
449640UK00009B/96